Cholesterol Transporters of the START Domain Protein Family in Health and Disease

Barbara J. Clark • Douglas M. Stocco
Editors

Cholesterol Transporters of the START Domain Protein Family in Health and Disease

START Proteins—Structure and Function

Springer

Editors
Barbara J. Clark, Ph.D.
Biochemistry and Molecular Biology
University of Louisville
School of Medicine
Louisville
Kentucky
USA

Douglas M. Stocco, Ph.D.
Cell Biology and Biochemistry
Texas Tech University
Health Sciences Center
Lubbock
Texas
USA

ISBN 978-1-4939-1111-0 ISBN 978-1-4939-1112-7 (eBook)
DOI 10.1007/978-1-4939-1112-7
Springer New York Heidelberg Dordrecht London

Library of Congress Control Number: 2014944308

Printed on acid-free paper

Springer is part of Springer Science+Business Media (www.springer.com)

The editors would like to dedicate this book to the memory of Dr. Keith L. Parker. Keith, whom many in the field of Endocrinology knew, respected, admired and loved, who passed away much too prematurely on December 13, 2008 at the age of 54. Keith was born in St. Louis, Missouri where he attended school through his high school years. He then enrolled at Williams College in Massachusetts and, while there, was elected to the Phi Beta Kappa honor society. From there Keith went to Washington University in St. Louis where he obtained M.D. and Ph.D. degrees in 1981. Keith then performed his internship and residency in Internal Medicine at Parkland Memorial Hospital in Dallas followed by a Fellowship in Genetics at Harvard Medical School. Upon leaving Harvard in 1986 he accepted his first position at Duke University Medical Center where he was also a Howard Hughes Investigator. Keith returned to Dallas in 1996 where he was the J. D. and Maggie E. Wilson Distinguished Professor in Biomedical Research at the University of Texas Southwestern Medical Center and remained at UTSW until his untimely death.

Keith was a scholar's scholar. He is best known for his pioneering work on SF1 that provided important and most interesting insight into the many facets of this indispensable transcription factor. The editors are especially indebted to Keith for his superb collaborations with us during the early stages of our own studies on steroidogenic acute regulatory protein (StAR). Keith's laboratory constructed the StAR knockout in mice, creating a model that was instrumental in corroborating the observations made on lipoid congenital adrenal hyperplasia in humans. He also contributed greatly to our understanding of the complexity with which the StAR gene is regulated. Keith will always be remembered for his brilliance, passion for science, generosity, humility, patience, humor and camaraderie. He was a true friend to many and is sorely missed to this day by all who knew him. We simply could not think of anyone who was more deserving of this dedication and hope that Keith will smile down upon us when he reads this.

Preface

It has been remarkable and rewarding to follow the explosion of work on the steroidogenic acute regulatory protein (StAR) and the classification of a new superfamily of proteins based on a shared StAR-related lipid-transfer (START) domain. Over the past 2 decades more than 1700 articles are tagged using "steroidogenic acute regulatory" as a search term in PubMed. This search captures articles on all START domain proteins, although the majority of the articles are on the expression, regulation and function of the STARD1 subfamily (StAR and STARD3/MLN64) with the cholesterol-binding START domain proteins of the STARD4 subfamily (STARD4/5/6) gaining attention. With 2014 being the 20th anniversary of the first report on StAR, our goals for this book are to present a compendium of the history and the current research on the STARD1 and STARD4 subfamily. Each chapter begins with a personal perspective of the discovery-to-publication journey that the authors had for the particular START domain family member that their laboratory identified or provided studies that quickly and significantly advanced our understanding of the function of these cholesterol transporters. One purpose for this unorthodox format for the scientific review articles herein is to give graduate students, post-doctoral fellows and endocrinology fellows a small glimpse of the research discovery process. We found the theme of collaboration and building upon previous great works is common to all stories.

Acknowledgements

The work summarized within the chapters of this book represents an ongoing story, a story that began over 6 decades ago. Therefore, we would like to thank the early contributors whose elegant work defined the acute regulation of steroidogenesis, and the current investigators in endocrinology laboratories worldwide devoted to discovery within this important research field. This book is only possible due to the research within the contributors' laboratories and their willingness to support our efforts on this project. We thank each of you; Pierre, Jean-Guy, Walt, Ann, Catherine, Ray, Fred and all of your respective co-authors for agreeing to be part of this story. Lastly, we would like to acknowledge people in our own laboratories who made our own work possible and fun. Thanks to Rebecca Combs, Ruth Wooton-Kee, Brian F. Clem, Renate Meier, YuChyu Chen, Elizabeth Hudson, Michael Kilgore, Jeffrey Wells, Debbie Alberts, Steven King, XingJia Wang, Zhiming Liu, Lance Walsh, Adam Reinhart, Matthew Dyson, Pulak Manna, Youngah Jo, Darrell Eubank, Rekha Rao, and Mariusz Kowalewski. We would like to acknowledge funds from NIH grant HD-17481 and the Robert A. Welch Foundation Grant B1-0028 that were instrumental in supporting the work in Dr. Stocco's lab and funds from NIH grant DK-51656, American Heart Association GIA-0051577B and 0555174B, and the University of Louisville Intramural Research programs for supporting the work in Dr. Clark's lab.

Contents

Contributors

Fabien Alpy Department of Functional Genomics and Cancer, Institut de Génétique et de Biologie Moléculaire et Cellulaire (IGBMC), Institut National de la Santé et de la Recherche Médicale (INSERM), U 964—Centre National de la Recherche Scientifique (CNRS) UMR 7104—University of Strasbourg, Illkirch, France

Faye Borthwick Metabolic and Cardiovascular Diseases Laboratory, Alberta Diabetes Institute, University of Alberta, Edmonton, AB, Canada

Barbara J. Clark Department of Biochemistry & Molecular Biology, University of Louisville, Louisville, KY, USA

Annette Graham Department of Life Sciences, School of Health and Life Sciences, Glasgow Caledonian University, Glasgow, UK

David B. Iaea Department of Biochemistry, Weill Cornell Medical College, New York, NY, USA

Pierre Lavigne Département de Biochimie, Faculté de médecine et des sciences de la santé, Université de Sherbrooke, Sherbrooke, QC, Canada

Andrée Lefebvre Département de Biochimie, Faculté de médecine et des sciences de la santé, Université de Sherbrooke, Sherbrooke, QC, Canada

Jean-Guy LeHoux Département de Biochimie, Faculté de médecine et des sciences de la santé, Université de Sherbrooke, Sherbrooke, QC, Canada

Danny Létourneau Département de Biochimie, Faculté de médecine et des sciences de la santé, Université de Sherbrooke, Sherbrooke, QC, Canada

Shu Mao Department of Biochemistry, Weill Cornell Medical College, New York, NY, USA

Frederick R. Maxfield Department of Biochemistry, Weill Cornell Medical College, New York, NY, USA

Walter L. Miller Department of Pediatrics, University of California San Francisco, San Francisco, CA, USA

Raymond E. Soccio Division of Endocrinology, Diabetes, and Metabolism, Smilow Center for Translational Research, Perelman School of Medicine, University of Pennsylvania, Philadelphia, PA, USA

Douglas M. Stocco Department of Cell Biology and Biochemistry, Texas Tech University Health Sciences Center, Lubbock, TX, USA

Janice Taylor Department of Life Sciences, School of Health and Life Sciences, Glasgow Caledonian University, Glasgow, UK

Catherine L. Tomasetto Department of Functional Genomics and Cancer, Institut de Génétique et de Biologie Moléculaire et Cellulaire (IGBMC), Institut National de la Santé et de la Recherche Médicale (INSERM), U 964—Centre National de la Recherche Scientifique (CNRS) UMR 7104—University of Strasbourg, Illkirch, France

Chapter 1
An Introduction to the Steroidogenic Acute Regulatory Protein (StAR)-Related Lipid Transfer Domain Protein Family

Barbara J. Clark and Douglas M. Stocco

Abstract The steroidogenic acute regulatory protein (StAR)-related lipid transfer (START) domain superfamily comprises a diverse group of proteins that bind hydrophobic lipids. The distinguishing feature shared by all members of this family is an α/β helix-grip fold structure containing a long hydrophobic pocket for ligand binding. The mammalian START domain protein family is grouped into 6 subfamilies that bind either cholesterol and oxysterols (STARD1/D3 and STARD4 subfamilies) or phospholipids and sphingolipids (STARD2/D11 subfamily), or have putative functions in Rho-GTPase signaling (STARD8/12/13 subfamily), thioesterase activity (STARD14/15 subfamily), or kinesin motor activity (STARD9). StAR (STARD1) is the namesake of the START domain protein family and has a well-established function in cholesterol transport in the adrenal and gonads for steroid hormone biosynthesis. Some of the mammalian START family members, e.g., STARD1, STARD11, and STARD2 are well characterized for their roles in cholesterol, ceramide, and phosphatidylcholine transfer, respectively, while much remains to be learned about the remaining family members. The purpose of this book is to present a compendium of the history of the discovery and the characterization of the mammalian START proteins, encompassing the seminal work over the past 50 years that has led to our current understanding of these lipid transport proteins. The chapters in this book focus on members of the STARD1/3 and STARD4 subfamilies, which have established roles involved in cholesterol and sterol trafficking. Each chapter provides a personal perspective of the discovery-to-publication journey for work on a START domain family member by authors whose work was instrumental in their discovery and characterization. This introductory chapter provides a brief overview and background on all members of the mammalian START protein family to provide a complete picture of this family of lipid transport proteins.

B. J. Clark, Ph.D. (✉)
Department of Biochemistry & Molecular Biology, University of Louisville,
Louisville, KY 40292 USA
e-mail: bjclark@louisville.edu

D. M. Stocco, Ph.D.
Department of Cell Biology and Biochemistry, Texas Tech University Health Sciences Center,
Lubbock, TX 79430 USA
e-mail: doug.stocco@ttuhsc.edu

B. J. Clark, D. M. Stocco (eds.), *Cholesterol Transporters of the START Domain Protein Family in Health and Disease,* DOI 10.1007/978-1-4939-1112-7_1,

Abbreviations

ACAT	acyl-CoA:cholesterol acyl transferase
ACOT	acyl-CoA thioesterase
BFIT2	brown fat-inducible thioesterase-2
CDCA	chenodeoxycholic acid
CERT	ceramide transfer protein
COL4A3BP	collagen type IV alpha 3 binding protein
DLC	deleted in liver cancer
FFAT	peptide EFFDAxE
FHA	forkhead-associated phosphopeptide binding domain
MENTAL	MLN64-N terminal domain
MLN64	metastatic axillary lymph node 64 kDa protein
NPC	Niemann –Pick type C disease
PC	phosphatidylcholine
PCTP	phosphatidylcholine transfer protein
PH	pleckstrin homology domain
SAM	sterile alpha domain
SREBP-2	sterol regulatory element binding protein-2
StAR	steroidogenic acute regulatory protein
START	StAR-related lipid transfer domain

Introduction

The StAR-related lipid transfer domain, abbreviated START, was first described by Ponting and Aravind [1] as a region of sequence similarity shared between a rat RhoGAP protein, plant Gla2 family members, mouse and human StAR, and bovine phosphatidylcholine transfer protein. Thus, the START domain was identified as a result of the testing of Web-based resources for the predictive value in identifying putative functional domains based on primary sequence data. The START domain is approximately 210 amino acids long and a distinguishing feature of this domain is the 3-D α/β helix-grip-fold structure defined by an antiparallel β-sheet flanked by amino- and carboxyl-terminal alpha helices [2]. The α/β helix-grip fold of the START domain proteins forms a U-shaped hydrophobic cleft that binds the ligand with the carboxyl-terminal alpha helix serving as a "cap" over the ligand binding cleft. Lipid access to the hydrophobic binding pocket requires a conformational change in the START domain and movement of the C-terminal helix [3, 4]. The helix-grip fold is used to define a large superfamily of proteins that bind hydrophobic lipids, classified as the SRPBCC[1] protein superfamily on NCBI's conserved domain database [5].

---

[1] NCBI cl14643: SRPBCC is the START/RHO_alpha_C/PITP/Bet_v1/CoxG/CalC superfamily. Rho_alpha_C, the C-terminal catalytic domains of the alpha oxygen-

START domains have been identified in plant, bacteria, protist, and animal genomes, with protein expression confirmed only in plant and animal species. Eighty percent of the START family is within the plant genome, in proteins that contain a homeodomain, suggesting a role in transcription [6]. Indeed, it is common to find START domains as part of multi-domain proteins that provide additional functions such as protein localization, enzymatic activity, or signaling [1, 2].

The mammalian START domain protein family is divided into six subfamilies based on sequence similarities [1, 7] (Table 1.1). In total, there are 15 members of the mammalian START protein family. Members of each subfamily share either similar ligand binding specificities or functional domains other than the START domain, such as the cholesterol and oxysterol binding proteins of the STARD1/D3 and STARD4/D5/D6 subfamilies, the phospholipid and sphingolipid binding proteins of the STARD2(PCTP)/D7/D10/D11 subfamily, the multi-domain proteins containing either putative Rho-GTPase signaling function of the STARD8/12/13 subfamily, thioesterase activity of the STARD14/15 subfamily, or kinesin motor function for STARD9. The first crystal structures for the START domains were reported for two mammalian START proteins, human STARD3/MLN64 and mouse STARD4 [8, 9]. Crystal structures for the START domains of hSTARD1, mSTARD4, hSTARD5, hSTARD2/PCTP, STARD11/CERT, hSTARD13, and hSTARD14 confirm the basic 3-D helix-grip fold structure across the five mammalian subfamilies [10–13]. One-third of the mammalian START domain proteins belong to the STARD1/D3 and STARD4 subfamilies and function to bind and transport cholesterol.

A brief background on the mammalian START domain family members is provided to complete the picture on these important lipid transporters (reviewed in [14, 15]). The remaining chapters of this book provide more detail on the regulation and function of members of the STARD1/D3 and STARD4 subfamilies.

The STARD1/STARD3 Subfamily

The STARD1 subfamily has two members, StAR (STARD1) and MLN64 (STARD3). StAR is the namesake of the START domain protein family and the deduced primary amino acid sequence was submitted to the NCBI database in November 1994 [16]. Shortly thereafter, a newly cloned and uncharacterized transcript from breast cancer, MLN64, was identified using differential screening of a complementary deoxyribonucleic acid (cDNA) library for amplified products in breast cancer-derived metastatic axillary lymph nodes (MLN). Protein database searches identified a domain within MLN64 that shared 33% sequence identity and 53% sequence similarity with the human StAR START domain [17, 18]. The significance

---

ase subunit of Rieske-type non-heme iron aromatic ring-hydroxylating oxygenases; PIPT, phosphatidylinositol transfer proteins; Bet v 1, the major pollen allergen of white birch, Betula verrucosa; CoxG, carbon monoxide dehydrogenase subunit G (gram-negative bacteria); CalC, and related proteins.

Table 1.1 Summary of the mammalian START domain family. The START protein subfamily is indicated and each member is designated by protein name followed by other known common name(s). A schematic of the domain structure highlights that the START domain is the C-terminal domain. Tissue distribution, cellular localization, lipid binding, and functional pathways and disease association are based on data from the cited references: Tissue distribution. [a]ubiquitous expression with the major tissues studied listed; [b]restricted expression [7, 19, 26, 28, 31, 33, 49, 50, 56, 59, 60, 69] cellular location; [c]motifs, domains direct subcellular location; [d]based on immunohistochemistry data for endogenous protein expression; [e]based on *in vitro* activity; [f]based on structure; lipid binding; [g]direct ligand binding assay; [h]modeled based on structure; [i]based on *in vitro* lipid extraction assay; [j]shown in crystal [9, 11–13, 29–31, 39, 43, 70–73]; function and disease association, [1][19]; [2][22]; [3][28]; [4][26, 74, 75]; [5][36]; [6][47]; [7–9][56]; [10,11][59] [12][42], [13][66]

START Subfamily	START protein	Domain structure	Tissue distribution[*,†]	Cellular location	Lipid binding	Metabolic pathway	Disease association
STARD1/D3	STARD1 (StAR)	SS–START	Adrenal, ovary, testis, brain[†], heart, liver	Mitochondria[‡,a,b,c]	Cholesterol[d,e]	steroidogenesis[1]	Lipoid CAH, HCC
	STARD3 (MLN64)	MENTAL-START	Placenta, breast, macrophages[*]	Transmembrane, late endosomes[‡,a,b,c]	Cholesterol[d,e]	endosomal cholesterol efflux[2]	NPC
STARD4	STARD4	START	Liver, macrophages, kidney[†]	Cytosolic[a] >> ER[a,b] mitochondria[b]	Cholesterol[¥,d]	ACAT activation[3]	
	STARD5	START	Macrophage, kidney proximal tubules[†]	Cytosolic >>ER, Golgi,PM[a]	CDCA[¥,d] Cholesterol, 25HC[¥,d]	ER stress response[4] ?	diabetic nephropathy
	STARD6	START	Testis germ cell[†]	Cytosolic[c], mitochondria[b]	Cholesterol[d,e]	?	
STARD2	STARD2 (PC-TP)	START	Liver, lung[*]	ER/Golgi[‡,b]	PC[f]	Glycolysis[5], FA synthesis[5]	insulin resistance
	STARD7 (GTT-1) STARD7-I	—START	Liver[*]	Cytosolic[c] mitochondria[c]	PC[e]	?	ovarian, lung, colon, liver cancer
	STARD10	—START—	Liver, kidney, testis, colon[*]	Cytosolic[c]	PC > PE[e]	bile acid conjugation/secretion[12]	fatty liver[12]
	STARD11 (CERT)	PH–FFAT–START	Liver[*]	ER/Golgi[‡,b]	Ceramide[¥,e]	ER -> Golgi ceramide transport[6]	
STARD8/12/13	STARD8 (DCL-3)	—RhoGAP–START	Cancer[*]	Focal adhesions[‡]	?	Tumor suppressor[7]	
	STARD12 (DCL-1)	—SAM–RhoGAP–START	Cancer[*]	Focal adhesions[‡] PM[a]	?	Cytoskeletal organization[8] Tumor suppressor8	liver cancer
	STARD13 (DCL-2)	—SAM–RhoGAP–START	Endothelial cells[*]	Focal adhesions[‡]	Charged lipid[d] (?)	Tumor suppressor[9]	angiogenesis
STARD9	STARD9	KM–FHA—CC–START	ubiquitous	nuclear	?	mitotic spindle formation[13]	mitosis
STARD14/15	STARD14 (ACOT11_v2, BFIT2)	Hotdog–Hotdog–START	Brown adipose tissue[†]	Cytosolic[c], ?	Fatty acid[d] (?)	Medium chain fatty-acyl-coA hydrolysis[10]	thermogenesis
	STARD15 (ACOT12)	–Hotdog–Hotdog–START	Liver[‡]	Cytosolic[c]	?	Acetyl-coA hydrolysis[11]	

of the START domain in MLN64 suggested a function in cholesterol binding and/or trafficking. Overall, the two members of the STARD1/D3 subfamily are similar in that the START domain for both proteins binds only cholesterol and additional sequences or domains localize the proteins to specific subcellular compartments. The differential subcellular localization of these START proteins suggests different functions in cholesterol trafficking (Table 1.1). First, StAR regulates cholesterol transfer into mitochondria and controls steroid hormone biosynthesis in the adrenal and gonads (reviewed in [19]). It is a nuclear-encoded protein that is synthesized in the cytosol as a 37 kDa precursor protein with an N-terminal mitochondria targeting sequence that directs the protein to the mitochondria. Mitochondrial import and processing of the precursor produces a 32 kDa intermediate product and a mature 30 kDa form that is localized within the matrix. The history and regulation of StAR in steroidogenic tissues is the topic for Chap. 2 while START domain structure and current models for cholesterol transport by StAR are the topics for Chaps. 3 and 4, respectively. Early studies indicated that StAR could transport cholesterol across mitochondrial membranes in many cell types, suggesting that StAR may function outside of steroidogenic tissues [20]. The action of StAR in non-steroidogenic cells is the topic for Chap. 5.

STARD3/MLN64 is a transmembrane protein that is targeted to the late endosomes. The location of STARD3/MLN64 to late endosomes led to studies on its potential role in Niemann Pick type C (NPC) disease, a lipid storage disorder caused by mutations in genes encoding either NPC1 or NPC2 that result in accumulation of cholesterol in lysosomes (reviewed in [21]). STARD3/MLN64 may function to shuttle cholesterol from NPC1 to a cytosolic acceptor protein or to an adjacent membrane [22–25]. STARD3 expression and function is the topic for Chap. 6.

The STARD4 Subfamily

STARD4 was first described as part of a gene set that was downregulated in mouse liver as a consequence of a high cholesterol diet. Five of the six genes that were identified with >2-fold decreased expression after high cholesterol diet were established cholesterol-regulated genes, and one was an uncharacterized EST (expressed sequence tag; [7]). Subsequent sequence and cloning studies confirmed that the EST was part of a transcript that contained a START domain; in fact the deduced amino acid sequence of the open reading frame encoded a protein of 224 amino acids composed entirely of the START domain. This EST was named STARD4.

Fifteen START domain-containing genes were identified in the human genome, 12 of which were previously characterized and two that were uncharacterized yet had 26–32 % sequence identity with STARD4. Based on the high sequence identity with STARD4, these genes were named sequentially STARD5 and STARD6 and formed the three members of the STARD4 subfamily [7]. The STARD4 family members represent ~22 kDa soluble proteins with no membrane targeting/association predicted (Table 1.1). STARD4 and STARD5 transcripts are abundant in liver and kidney tissues. In the liver, STARD4 is found in the hepatocytes and Kupffer cells, the macrophages of the liver, while STARD5 is detected only in Kupffer cells. In kidney, both STARD4 and STARD5 protein are present in the proximal tubules [26].

STARD4 binds only cholesterol with a proposed biological role for STARD4 in regulating cholesterol sensing by the endoplasmic reticulum and cholesterol ester synthesis [27–29]. STARD5 binds primary bile acids, specifically chenodeoxycholic acid, with high affinity [30]. The fact that STARD5 is a bile acid binding/transporting START protein was a surprise to the field given that initial reports suggested cholesterol and 25-hydroxycholesterol were the ligands for STARD5 [7, 27, 29, 31]. The potential biological functions for StARD5 appear to be different from those of STARD4, in that STARD5 expression is not linked to increased cholesterol ester levels but rather to increased free cholesterol levels [26, 29, 31, 32]. It remains to be determined whether STARD5 plays a role in cholesterol homeostasis, yet it is clear that both direct and indirect mechanisms should be explored. For example, STARD5 may modulate the activity of the bile acid-activated nuclear receptor farnesol-X-receptor (FXR) by either promoting or blocking ligand binding. FXR controls expression of genes involved in cholesterol, bile acid, and lipid homeostasis, thus, STARD5 could potentially indirectly regulate FXR-dependent signaling. Chapters 7 and 8 present the seminal work on STARD4 and STARD5 expression, regulation, and function, and explores further the current questions on the role of STARD5 in cholesterol homeostasis.

STARD6 is expressed in mouse testis, specifically in the germ cells with highest expression in round spermatids [7, 33]. The function of STARD6 in spermatogenesis is not known, although it may play a role in mitochondrial NADH-dependent dehydrogenase activity (diaphorase) associated with sperm motility and quality [34].

The STARD2/Phosphatidylcholine Transfer Protein (PCTP) Subfamily

STARD2, STARD7, and STARD10 all bind phosphatidylcholine (PC) with STARD2 being the best characterized PC transporter [12]. The early work on STARD2, also named phosphatidylcholine transfer protein (PCTP), tested its effect on PC transport in the liver and lung, reasoning that the protein is highly expressed in these tissues where PC is a major phospholipid in bile and lung surfactant [35]. The direct test was to generate mice deficient in STARD2/PCTP (*Pctp*−/−) and measure PC in the bile and lung surfactant [35]. The unpredicted result was the *Pctp*−/− mice have the same bile or lung surfactant PC content as their wild-type counterparts. However, the investigators continued to characterize these mice and made an interesting observation that fasting serum glucose and free fatty acids levels were significantly decreased in *Pctp*−/− mice [36]. Similarly, inhibiting PC binding to STARD2/PCTP in wild-type mice decreased the effects of a high fat diet on serum glucose levels [37]. Blocking STARD2/PCTP action, i.e., PC binding in cultured human hepatocytes activated the insulin signaling pathway [37]. Therefore, it appears that STARD2/PCTP functions in the liver to suppress insulin sensitivity. This unpredicted role for STARD2 opens new possibilities for targeting STARD2 in treatment of diabetes. Future studies are necessary to determine the mechanism for STARD2/PCTP action in liver glucose metabolism.

STARD7 also binds PC specifically and shares 25% sequence identity with STARD2/PCTP [38, 39]. The protein has been detected in lung, colon, and liver cancer cell lines [38], and therefore, has tissue overlap with STARD2. A STARD7 variant, STARD7-1, with an amino terminal mitochondrial targeting sequence has been associated with increased mitochondrial PC levels when overexpressed in a cultured mouse hepatoma cell line [39]. Both cytosolic and mitochondrial localization of endogenous STARD7-I in liver cells and tissue has been demonstrated, consistent with the increased PC levels in mitochondria with overexpression of the protein. Currently, it is proposed that STARD7 may play a role in mitochondrial biogenesis, yet this has not been directly addressed [39].

STARD10 expression has been reported in mouse testicular germ cells, liver, intestine, and human mammary tissue [40–42]. PC and phosphatidylethanolamine both bind STARD10 although PC binds with greater affinity [40, 43]. The PC transfer activity of STARD10 appears to be regulated by posttranslational modification; phosphorylation of STARD10 decreases membrane association and PC extraction [44]. A putative role in breast cancer has been proposed with STARD10 functioning within the ErbB2/HER2/neu receptor signaling pathway, although the coordinated expression of STARD10 and HER2/neu in breast tumors may not be directly associated [40, 45]. Additionally, loss of STARD10 expression was found to be an independent marker for poor patient outcome and may be used to identify a specific subgroup of patients at high risk [45]. Presently, the function of STARD10 in mammary tissues is not defined so the question remains whether it contributes to the anti- or pro-oncogenic breast tumor phenotype. The newly described *Stard10* knockout mice (Stard10$^{-/-}$) may serve as a good model to address this question directly [42]. Interestingly, characterization of the Stard10$^{-/-}$ mice provided some unexpected results; the levels of PC in the liver and bile were not different between the wild-type and the Stard10$^{-/-}$ mice, rather the major disorder was in bile acid metabolism with an increase in biliary secretion of conjugated bile acids. Much of the change in bile acid metabolism is attributed to the peroxisome proliferator-activated receptor-alpha (PPARα)-mediated changes in gene expression in the liver and intestine [42].

STARD11 is more commonly referred to as CERT, for ceramide transport protein. STARD11/CERT is responsible for the movement of ceramide from the ER to the Golgi membrane [46]. CERT is a multi-domain protein containing an amino terminal pleckstrin homology domain (PH), a FFAT motif, and carboxyl-terminal START domain. The domain structure of CERT orients the protein between ER and Golgi membranes; the PH domain binds to the phosphoinositide PI4P in the Golgi membrane and the FFAT motif binds to a vesicle-associated membrane protein-associated protein (VAP), an integral ER membrane protein. This orientation of the protein models the START domain in close proximity to both membranes so that association with the ER extracts ceramide from the membrane and association with the Golgi promotes ceramide absorption by the membrane [10, 11, 47]. Within this subfamily, STARD2/PCTP and STARD11/CERT have been studied in greater detail and there are several reviews on their potential functions [47–50].

The STARD8/12/13 Subfamily

The members of this subfamily contain a sterile alpha motif [13], a serine-rich region, a RhoGAP domain, and a START domain that make this subfamily the most complex in terms of domain structure ([13] Table 1.1). STARD12, the first member identified within this family, is contained in a genomic region that is associated with loss of heterozygosity in several cancers, hence, members of this subfamily are also referred to as the *d*eleted in *l*iver *c*ancer (DLC) family of proteins [51]. STARD12/DLC-1, STARD13/DLC-2 and STARD8/DLC-3 all function as tumor suppressors based on studies that showed increased apoptosis and decreased cell growth in several cancer cell lines after re-expression of one of the proteins [52, 53]. The basis for the tumor suppressor function is not known, yet the multi-domain nature of these proteins and the subcellular localization provide clues to function and suggest a potential for integrated functions. For example, imaging studies in live cells have shown that the STARD12/13/8 proteins are present at the plasma membrane, at regions associated with cytoskeletal proteins (focal adhesions) and lipid rafts enriched in cholesterol and sphingolipid [54, 55]. The lipid that binds the START domain in this family has not been determined, although the crystal structure for STARD13/DLC-2 predicts the ligand binding pocket would accommodate a charged lipid rather than cholesterol or a phospholipid [13]. The location of these proteins at the plasma membrane leads to speculation that ligand binding in the START domain may regulate the RhoGAP function, thereby controlling Rho signaling and cell proliferation [55, 56].

The STARD14/15 Subfamily

STARD14 and STARD15 contain an ACOT (acyl-coenzyme A thioesterase) domain in addition to the START domain and functions in the hydrolysis of coenzyme A [57] from activated fatty acids and acetyl-CoA [58, 59]. Human STARD14 is the ortholog of the mouse brown fat-inducible thioesterase (mBFIT2; [60] Table 1.1) and its expression is associated with increased metabolic activity. STARD15/ACOT12 is a cytosolic acetyl-CoA thioesterase (hydrolase) that is highly expressed in liver [61–63]. The crystal structure for the START domain of STARD14 predicts [13] a fatty acid would fill the ligand-binding cavity. Although neither the ligand nor role of the START domain is known for this subfamily, the evolution of the two domain structure would indicate a convergence of purpose. One model is that fatty acid binding in the START domain regulates the thioesterase activity, similar to that proposed for the START domain-RhoGAP interaction of the STARD8/12/13 subfamily.

The STARD9 Subfamily

The STARD9 subfamily contains a single member which has an amino-to-carboxy terminal multi-domain structure composed of a kinesin motor domain, a forkhead-associated phosphopeptide binding domain (FHA) and the START domain [7, 64, 65]. The STARD9 kinesin and FHA domains have sequence similarity (~50% identity) to the kinesin-3 family members KIF16B and KIF1A. STARD9 was identified as a protein that associated with mitotic microtubules and the amino terminal kinesin domain of STARD9 is active for microtubule binding and adenosine triphosphate (ATP) hydrolysis, supporting STARD9 functions as a kinesin motor protein. Small interfering ribonucleic acid (siRNA) depletion of STARD9 in several cancer cell lines revealed the protein is important for microtubule spindle assembly and mitosis; loss of STARD9 resulted in increased mitotic arrest and increased apoptosis [66]. The loss of STARD9 also enhanced the sensitivity of the cells to anti-mitotic cancer drugs, implicating STARD9 as a putative novel target for cancer treatment. However, both the identity of the lipid bound by the START domain and the function of lipid binding for STARD9 function remains to be determined.

Summary

The START domain family is composed of lipid transport proteins that function in the non-vesicular trafficking of sterols and phospho/sphingolipids. The STARD1/D3 subfamily members are established cholesterol binding/transport proteins, yet defining the mechanisms of sterol transport for these proteins remains a current challenge. StAR, the START protein family namesake, regulates cholesterol transport across mitochondrial membranes for steroid hormone synthesis and defining the mechanism(s) for StAR-dependent cholesterol transport has been investigated using both structural and functional approaches. Current models for mitochondrial cholesterol transport by StAR are presented in Chaps. 3 and 4 while the potential for an expanded role for StAR in non-steroidogenic tissues is presented in Chap. 5. STARD3 is a membrane protein localized to the late endosome and has a putative role in shuttling cholesterol derived from exogenous sources to cytosolic cholesterol carriers for subsequent cellular distribution. Chapter 6 provides the discovery-to-current models for function journey for STARD3.

The tissue distribution profiles, gene regulation, and ligand specificities suggest unique functions for the STARD4 family members. As soluble lipid transport proteins, this subfamily is proposed to participate in the non-vesicular trafficking of cholesterol between biological membranes to help maintain the proper cholesterol: phospholipid:sphingolipid distribution (reviewed in [67, 68]). Most of the work reported has been on STARD4 and the regulation and function of STARD4 is the topic for Chaps. 7 and 8 while a general overview of non-vesicular cholesterol trafficking and the role of the STARD4 subfamily in this process is the topic for Chap. 8.

Functions described for members of the remaining START subfamilies reveal a very diverse role for this family of proteins. The knockout mouse models for members of the STARD2 phospholipid/sphingolipid binding proteins have provided unexpected and interesting results that indicate potential broad applications for this subfamily in regulating glucose metabolism in diabetes (STARD2), tumor proliferation in cancer (STARD7/10), bile acid metabolism in the liver and intestine (STARD10), and ceramide transfer from the ER to the Golgi membrane for sphingolipid synthesis (STARD11/CERT). Although PC phospholipid metabolism or trafficking appears to be normal in the absence of STARD2 or STARD10, a remaining question is whether lipid binding to the START domain impacts the function of these proteins. The lipid specificity for the remaining START proteins in the RhoGAP multi-domain subfamily, the thioesterase subfamily, and the STARD9 subfamily has not been determined. The RhoGAP multidomain START proteins are defined by their loss in cancer and can function as tumor suppressors although this function is attributed to the RhoGAP domain. The thioesterase START protein subfamily functions in fatty acid hydrolysis while STARD9 binds to centromeres and is important for mitotic spindle formation, again both functions attributed to other domains within the proteins. It is plausible to predict that lipid binding in the START domains will regulate the described function for these proteins, and testing this prediction will require identifying the lipid ligands.

Lastly, as disease states can provide significant insight into biological function, START protein expression and function should be considered when examining disorders that involve dyslipidemia and inflammation. Since dyslipidemia and inflammation are common to obesity, diabetes, coronary heart disease, and cancer, there is much to be learned about this family of lipid transport proteins.

References

1. Ponting CP, Aravind L. START: a lipid-binding domain in StAR, HD-ZIP and signalling proteins. Trends Biochem Sci. 1999;24:130–2.
2. Iyer LM, Koonin EV, Aravind L. Adaptations of the helix-grip fold for ligand binding and catalysis in the START domain superfamily. Proteins. 2001;43:134–44.
3. Baker BY, Yaworsky DC, Miller WL. A pH-dependent molten globule transition is required for activity of the steroidogenic acute regulatory protein, StAR. J Biol Chem. 2005;280:41753–60.
4. Bose HS, Whittal RM, Ran Y, Bose M, Baker BY, Miller WL. StAR-like activity and molten globule behavior of StARD6, a male germ-line protein. Biochemistry. 2008;47:2277–88.
5. Marchler-Bauer A, Anderson JB, Chitsaz F, Derbyshire MK, DeWeese-Scott C, Fong JH, Geer LY, Geer RC, Gonzales NR, Gwadz M, He S, Hurwitz DI, Jackson JD, Ke Z, Lanczycki CJ, Liebert CA, Liu C, Lu F, Lu S, Marchler GH, Mullokandov M, Song JS, Tasneem A, Thanki N, Yamashita RA, Zhang D, Zhang N, Bryant S. CDD: specific functional annotation with the Conserved Domain Database. Nucleic Acids Res. 2009;37:D205–10.

6. Schrick K, Nguyen D, Karlowski WM, Mayer KF. START lipid/sterol-binding domains are amplified in plants and are predominantly associated with homeodomain transcription factors. Genome Biol. 2004;5:R41.
7. Soccio RE, Adams RM, Romanowski MJ, Sehayek E, Burley SK, Breslow JL. The cholesterol-regulated StarD4 gene encodes a StAR-related lipid transfer protein with two closely related homologues, StarD5 and StarD6. Proc Natl Acad Sci U S A. 2002;99:6943–8.
8. Romanowski MJ, Soccio RE, Breslow JL, Burley S. Crystal structure of the Mus musculus cholesterol-regulated START protein 4 (StarD4) containing a StAR-related lipid transfer domain. Proc Natl Acad Sci U S A. 2002;99:6949–54.
9. Tsujishita Y, Hurley JH. Structure and lipid transport mechanism of a StAR-related domain. Nat Struct Biol. 2000;7:408–14.
10. Kudo N, Kumagai K, Matsubara R, Kobayashi S, Hanada K, Wakatsuki S, Kato R. Crystal structures of the CERT START domain with inhibitors provide insights into the mechanism of ceramide transfer. J Mol Biol. 2010;396:245–51.
11. Kudo N, Kumagai K, Tomishige N, Yamaji T, Wakatsuki S, Nishijima M, Hanada K, Kato R. Structural basis for specific lipid recognition by CERT responsible for nonvesicular trafficking of ceramide. Proc Natl Acad Sci U S A. 2008;105:488–93.
12. Roderick SL, Chan WW, Agate DS, Olsen LR, Vetting MW, Rajashankar KR, Cohen DE. Structure of human phosphatidylcholine transfer protein in complex with its ligand. Nat Struct Biol. 2002;9:507–11.
13. Thorsell AG, Lee WH, Persson C, Siponen MI, Nilsson M, Busam RD, Kotenyova T, Schuler H, Lehtio L. Comparative structural analysis of lipid binding START domains. PLoS One. 2011;6:e19521.
14. Clark BJ. The mammalian START domain protein family in lipid transport in health and disease. J Endocrinol. 2012;212:257–75.
15. Alpy F, Tomasetto C. START ships lipids across interorganelle space. Biochimie. 2013;96:85–95.
16. Clark BJ, Wells J, King SR, Stocco DM. The purification, cloning, and expression of a novel luteinizing hormone-induced mitochondrial protein in MA-10 mouse Leydig tumor cells. Characterization of the steroidogenic acute regulatory protein (StAR). J Biol Chem. 1994;269:28314–22.
17. Moog-Lutz C, Tomasetto C, Regnier CH, Wendling C, Lutz Y, Muller D, Chenard MP, Basset P, Rio MC. MLN64 exhibits homology with the steroidogenic acute regulatory protein (STAR) and is over-expressed in human breast carcinomas. Int J Cancer. 1997;71:183–91.
18. Tomasetto C, Regnier C, Moog-Lutz C, Mattei MG, Chenard MP, Lidereau R, Basset P, Rio MC. Identification of four novel human genes amplified and overexpressed in breast carcinoma and localized to the q11–q21.3 region of chromosome 17. Genomics. 1995;28:367–76.
19. Stocco DM. StAR protein and the regulation of steroid hormone biosynthesis. Ann Rev Physiol. 2001;63:193–213.
20. Sugawara T, Lin D, Holt JA, Martin KO, Javitt NB, Miller WL, Strauss JF, 3rd. Structure of the human steroidogenic acute regulatory protein (StAR) gene: StAR stimulates mitochondrial cholesterol 27-hydroxylase activity. Biochemistry. 1995;34:12506–12.
21. Rosenbaum AI, Maxfield FR. Niemann-Pick type C disease: molecular mechanisms and potential therapeutic approaches. J Neurochem. 2011;116:789–95.
22. Alpy F, Tomasetto C. MLN64 and MENTHO, two mediators of endosomal cholesterol transport. Biochem Soc Trans. 2006;34:343–5.
23. Alpy F, Latchumanan VK, Kedinger V, Janoshazi A, Thiele C, Wendling C, Rio MC, Tomasetto C. Functional characterization of the MENTAL domain. J Biol Chem. 2005;280:17945–52.
24. Charman M, Kennedy BE, Osborne N, Karten B. MLN64 mediates egress of cholesterol from endosomes to mitochondria in the absence of functional Niemann-Pick Type C1 protein. J Lipid Res. 2009;51:1023–34.
25. Alpy F, Rousseau A, Schwab Y, Legueux F, Stoll I, Wendling C, Spiegelhalter C, Kessler P, Mathelin C, Rio MC, Levine TP, Tomasetto C. STARD3/STARD3NL and VAP make a novel molecular tether between late endosomes and the ER. J Cell Sci. 2013;126:5500–12.

26. Chen Y-C, Meier RK, Zheng S, Khundmiri SJ, Tseng MT, Lederer ED, Epstein PN, Clark BJ. Steroidogenic Acute Regulatory (StAR)-Related Lipid Transfer Domain Protein 5 (STARD5) localization and regulation in renal tubules. Am J Physiol Renal Physiol. 2009;297:F380–8. doi:90433.92008.
27. Mesmin B, Pipalia NH, Lund FW, Ramlall TF, Sokolov A, Eliezer D, Maxfield FR. STARD4 abundance regulates sterol transport and sensing. Mol Biol Cell. 2011;22:4004–15. doi:10.1091/mbc.E11-04-0372.
28. Rodriguez-Agudo D, Calderon-Dominguez M, Ren S, Marques D, Redford K, Medina-Torres MA, Hylemon P, Gil G, Pandak WM. Subcellular localization and regulation of StarD4 protein in macrophages and fibroblasts. Biochim Biophys Acta. 2011;1811:597–606.
29. Rodriguez-Agudo D, Ren S, Wong E, Marques D, Redford K, Gil G, Hylemon P, Pandak WM. Intracellular cholesterol transporter StarD4 binds free cholesterol and increases cholesteryl ester formation. J Lipid Res. 2008;49:1409–19.
30. Letourneau D, Lorin A, Lefebvre A, Frappier V, Gaudreault F, Najmanovich R, Lavigne P, LeHoux JG. StAR-related lipid transfer domain protein 5 binds primary bile acids. J Lipid Res. 2012;53:2677–89.
31. Rodriguez-Agudo D, Ren S, Hylemon PB, Redford K, Natarajan R, Del Castillo A, Gil G, Pandak WM. Human StarD5, a cytosolic StAR-related lipid binding protein. J Lipid Res. 2005;46:1615–23.
32. Borthwick F, Allen AM, Taylor JM, Graham A. Overexpression of STARD3 in human monocyte/macrophages induces an anti-atherogenic lipid phenotype. Clin Sci (Lond). 2010;119:265–72.
33. Gomes C, Oh SD, Kim JW, Chun SY, Lee K, Kwon HB, Soh J. Expression of the putative sterol binding protein Stard6 gene is male germ cell specific. Biol Reprod. 2005;72:651–8.
34. Golas A, Malek P, Piasecka M, Styrna J. Sperm mitochondria diaphorase activity-a gene mapping study of recombinant inbred strains of mice. Int J Dev Biol. 2010;54:667–73.
35. van Helvoort A, de Brouwer, Elferink RP, Borst P. Mice without phosphatidylcholine transfer protein have no defects in the secretion of phosphatidylcholine into bile or into lung airspaces. Proc Natl Acad Sci U S A. 1999;96:11501–6.
36. Scapa EF, Pocai A, Wu MK, Gutierrez-Juarez R, Glenz L, Kanno K, Li H, Biddinger S, Jelicks LA, Rossetti L, Cohen DE. Regulation of energy substrate utilization and hepatic insulin sensitivity by phosphatidylcholine transfer protein/StarD2. FASEB J. 2008;22:2579–90.
37. Shishova EY, Stoll JM, Ersoy BA, Shrestha S, Scapa EF, Li Y, Niepel MW, Su Y, Jelicks LA, Stahl GL, Glicksman MA, Gutierrez-Juarez R, Cuny GD, Cohen DE. Genetic ablation or chemical inhibition of phosphatidylcholine transfer protein attenuates diet-induced hepatic glucose production. Hepatology. 2011;54:664–74.
38. Durand S, Angeletti S, Genti-Raimondi S. GTT1/StarD7, a novel phosphatidylcholine transfer protein-like highly expressed in gestational trophoblastic tumour: cloning and characterization. Placenta. 2004;25:37–44.
39. Horibata Y, Sugimoto H. StarD7 mediates the intracellular trafficking of phosphatidylcholine to mitochondria. J Biol Chem. 2010;285:7358–65.
40. Olayioye MA, Hoffmann P, Pomorski T, Armes J, Simpson RJ, Kemp BE, Lindeman GJ, Visvader JE. The phosphoprotein StarD10 is overexpressed in breast cancer and cooperates with ErbB receptors in cellular transformation. Cancer Res. 2004;64:3538–44.
41. Yamanaka M, Koga M, Tanaka H, Nakamura Y, Ohta H, Yomogida K, Tsuchida J, Iguchi N, Nojima H, Nozaki M, Matsumiya K, Okuyama A, Toshimori K, Nishimune Y. Molecular cloning and characterization of phosphatidylcholine transfer protein-like protein gene expressed in murine haploid germ cells. Biol Reprod. 2000;62:1694–1701.
42. Ito M, Yamanashi Y, Toyoda Y, Izumi-Nakaseko H, Oda S, Sugiyama A, Kuroda M, Suzuki H, Takada T, Adachi-Akahane S. Disruption of Stard10 gene alters the PPARα-mediated bile acid homeostasis. Biochim Biophys Acta (BBA)—Mol Cell Biol Lipids. 2013;1831:459–68.

43. Olayioye MA, Vehring S, Muller P, Herrmann A, Schiller J, Thiele C, Lindeman GJ, Visvader JE, Pomorski T. StarD10, a START domain protein overexpressed in breast cancer, functions as a phospholipid transfer protein. J Biol Chem. 2005;280:27436–42.
44. Olayioye MA, Buchholz M, Schmid S, Schoffler P, Hoffmann P, Pomorski T. Phosphorylation of StarD10 on serine 284 by casein kinase II modulates its lipid transfer activity. J Biol Chem. 2007;282:22492–8.
45. Murphy NC, Biankin AV, Millar EK, McNeil CM, O'Toole SA, Segara D, Crea P, Olayioye MA, Lee CS, Fox SB, Morey AL, Christie M, Musgrove EA, Daly RJ, Lindeman GJ, Henshall SM, Visvader JE, Sutherland RL. Loss of STARD10 expression identifies a group of poor prognosis breast cancers independent of HER2/Neu and triple negative status. Int J Cancer. 2009;126:1445–53.
46. Hanada K, Kumagai K, Yasuda S, Miura Y, Kawano M, Fukasawa M, Nishijima M. Molecular machinery for non-vesicular trafficking of ceramide. Nature. 2003;426:803–9.
47. Hanada K, Kumagai K, Tomishige N, Yamaji T. CERT-mediated trafficking of ceramide. Biochim Biophys Acta. 2009;1791:684–91.
48. Kang HW, Wei J, Cohen DE. PC-TP/StARD2: of membranes and metabolism. Trends Endocrinol Metab. 2010;21:449–56.
49. Mencarelli C, Losen M, Hammels C, De Vry J, Hesselink MK, Steinbusch HW, De Baets MH, Martinez-Martinez P. The ceramide transporter and the Goodpasture antigen binding protein: one protein-one function? J Neurochem. 2010;113:1369–86.
50. Kanno K, Wu MK, Scapa EF, Roderick SL, Cohen DE. Structure and function of phosphatidylcholine transfer protein (PC-TP)/StarD2. Biochim Biophys Acta. 2007;1771:654–62.
51. Yuan BZ, Miller MJ, Keck CL, Zimonjic DB, Thorgeirsson SS, Popescu NC. Cloning, characterization, and chromosomal localization of a gene frequently deleted in human liver cancer (DLC-1) homologous to rat RhoGAP. Cancer Res. 1998;58:2196–9.
52. Ching YP, Wong CM, Chan SF, Leung TH, Ng DC, Jin DY, Ng IO. Deleted in liver cancer (DLC) 2 encodes a RhoGAP protein with growth suppressor function and is underexpressed in hepatocellular carcinoma. J Biol Chem. 2003;278:10824–30.
53. Durkin ME, Ullmannova V, Guan M, Popescu NC. Deleted in liver cancer 3 (DLC-3), a novel Rho GTPase-activating protein, is downregulated in cancer and inhibits tumor cell growth. Oncogene. 2007;26:4580–9.
54. Kim TY, Healy KD, Der CJ, Sciaky N, Bang YJ, Juliano RL. Effects of structure of Rho GTPase-activating protein DLC-1 on cell morphology and migration. J Biol Chem. 2008;283:32762–70.
55. Yamaga M, Sekimata M, Fujii M, Kawai K, Kamata H, Hirata H, Homma Y, Yagisawa H. A PLCdelta1-binding protein, p122/RhoGAP, is localized in caveolin-enriched membrane domains and regulates caveolin internalization. Genes Cells. 2004;9:25–37.
56. Durkin ME, Yuan BZ, Zhou X, Zimonjic DB, Lowy DR, Thorgeirsson SS, Popescu NC. DLC-1: a Rho GTPase-activating protein and tumour suppressor. J Cell Mol Med. 2007;11:1185–207.
57. Artemenko IP, Zhao D, Hales DB, Hales KH, Jefcoate CR. Mitochondrial processing of newly synthesized steroidogenic acute regulatory protein (StAR), but not total StAR, mediates cholesterol transfer to cytochrome P450 side chain cleavage enzyme in adrenal cells. J Biol Chem. 2001;276:46583–96.
58. Hunt MC, Yamada J, Maltais LJ, Wright MW, Podesta EJ, Alexson S. A revised nomenclature for mammalian acyl-CoA thioesterases/hydrolases. J Lipid Res. 2005;46:2029–32.
59. Kirkby B, Roman N, Kobe B, Kellie S, Forwood JK. Functional and structural properties of mammalian acyl-coenzyme A thioesterases. Prog Lipid Res. 2010;49:366–77.
60. Adams SH, Chui C, Schilbach SL, Yu XX, Goddard AD, Grimaldi JC, Lee J, Dowd P, Colman S, Lewin DA. BFIT, a unique acyl-CoA thioesterase induced in thermogenic brown adipose tissue: cloning, organization of the human gene and assessment of a potential link to obesity. Biochem J. 2001;360:135–42.
61. Suematsu N, Isohashi F. Molecular cloning and functional expression of human cytosolic acetyl-CoA hydrolase. Acta Biochim Pol. 2006;53:553–61.

62. Suematsu N, Okamoto K, Isohashi F. Mouse cytosolic acetyl-CoA hydrolase, a novel candidate for a key enzyme involved in fat metabolism: cDNA cloning, sequencing and functional expression. Acta Biochim Pol. 2002;49:937–45.
63. Suematsu N, Okamoto K, Shibata K, Nakanishi Y, Isohashi F. Molecular cloning and functional expression of rat liver cytosolic acetyl-CoA hydrolase. Eur J Biochem. 2001;268:2700–9.
64. Alpy F, Tomasetto C. Give lipids a START: the StAR-related lipid transfer (START) domain in mammals. J Cell Sci. 2005;118:2791–801.
65. Soccio RE, Breslow JL. StAR-related lipid transfer (START) proteins: mediators of intracellular lipid metabolism. J Biol Chem. 2003;278:22183–6.
66. Torres Jorge Z, Summers Matthew K, Peterson D, Brauer Matthew J, Lee J, Senese S, Gholkar Ankur A, Lo Y-C, Lei X, Jung K, Anderson David C, Davis David P, Belmont L, Jackson Peter K. The STARD9/Kif16a kinesin associates with mitotic microtubules and regulates spindle pole assembly. Cell. 2011;147:1309–23.
67. Lev S. Non-vesicular lipid transport by lipid-transfer proteins and beyond. Nat Rev Mol Cell Biol. 2010;11:739–50.
68. Prinz WA. Non-vesicular sterol transport in cells. Prog Lipid Res. 2007;46:297–314.
69. Strauss JF, 3rd, Kishida T, Christenson LK, Fujimoto T, Hiroi H. START domain proteins and the intracellular trafficking of cholesterol in steroidogenic cells. Mol Cell Endocrinol. 2003;202:59–65.
70. Barbar E, Lehoux JG, Lavigne P. Toward the NMR structure of StAR. Mol Cell Endocrinol. 2009;300:89–93.
71. Bose M, Whittal RM, Miller WL, Bose HS. Steroidogenic activity of StAR requires contact with mitochondrial VDAC1 and phosphate carrier protein. J Biol Chem. 2008;283:8837–45.
72. Murcia M, Faraldo-Gomez JD, Maxfield FR, Roux B. Modeling the structure of the StART domains of MLN64 and StAR proteins in complex with cholesterol. J Lipid Res. 2006;47:2614–30.
73. Lorin A, Letourneau D, Lefebvre A, LeHoux JG, Lavigne P. (1)H, (13)C, and (15)N backbone chemical shift assignments of StAR-related lipid transfer domain protein 5 (STARD5). Biomol NMR Assign. 2013;7:21–4.
74. Soccio RE, Adams RM, Maxwell KN, Breslow JL. Differential gene regulation of StarD4 and StarD5 cholesterol transfer proteins: activation of StARD4 by sterol regulatory element-binding protein-2 and StARD5 by endoplasmic reticulum stress. J Biol Chem. 2005;280:19410–8.
75. Rodriguez-Agudo D, Calderon-Dominguez M, Medina MA, Ren S, Gil G, Pandak WM. ER stress increases StarD5 expression by stabilizing its mRNA and leads to relocalization of its protein from the nucleus to the membranes. J Lipid Res. 2012;53:2708–15.

Chapter 2
The Steroidogenic Acute Regulatory Protein (StAR)

Barbara J. Clark and Douglas M. Stocco

Abstract StAR, the Steroidogenic Acute Regulatory protein, was named for its critical role in the acute regulation of steroid hormone biosynthesis in the adrenal and gonads following tropic hormone stimulation. StAR synthesis is required for the first and rate-limiting step in steroid hormone biosynthesis, cholesterol transport into mitochondria. It was a long journey to finding the acute regulator of steroidogenesis, and this chapter provides a historical and personal account of this journey from the perspective of Dr. Douglas M. Stocco. Over the past two decades, we have gained significant insight into the mechanisms that regulate StAR expression, and into StAR structure and function. This chapter also provides a summary of the literature that has led to our current understanding of the cyclic adenosine-3′,5′-monophosphate (cAMP)-protein kinase A-dependent mechanisms that control StAR expression at the transcriptional and post-transcriptional levels in steroidogenic tissues.

Abbreviations

AC	adenylyl cyclase
ACTH	adrenocorticotropic hormone
AMP	adenosine-3′,5′-monophosphate
AP-1	activator protein 1
AURE	adenosine-uridine-rich destabilizing element
Bt_2cAMP	N6,2′-O-*dibutyryl*-adenosine-3′,5′-monophosphate
bZIP	basic leucine zipper
cAMP	cyclic AMP
CBP/p300	CREB-binding protein

B. J. Clark, Ph.D. (✉)
Department of Biochemistry & Molecular Biology, University of Louisville,
Louisville, KY 40292, USA
e-mail: bjclark@louisville.edu

D. M. Stocco, Ph.D.
Department of Cell Biology and Biochemistry,
Texas Tech University Health Sciences Center, Lubbock, TX 79430, USA
e-mail: doug.stocco@ttuhsc.edu

B. J. Clark, D. M. Stocco (eds.), *Cholesterol Transporters of the START Domain Protein Family in Health and Disease,* DOI 10.1007/978-1-4939-1112-7_2,

C/EBPβ	CCAAT enhancer-binding protein-beta
COUP-TF	chicken ovalbumin upstream promoter transcription factor I and II
CREB	cyclic-AMP responsive element-binding protein
CREM	CREB modulator protein
CRH	corticotropin hormone
DAG	diacylglycerol
DAX-1	dosage sensitive sex reversal-adrenal hypoplasia congenital gene on the X chromosome gene 1
ER	endoplasmic reticulum
FSH	follicule stimulating hormone
GnRH	gonadotropin-releasing hormone
IP3	inositol 1, 4, 5 trisphosphate
LH	luteinizing hormone
NPC	Niemann Pick type C
NUR77/NGFI-B	nerve growth factor induced-B
PAP7	TSPO associated protein 7 (ACBD3)
PIC	preinitiation complex
PKA	protein kinase A
PKC	protein kinase C
Poly(A)	polyadenylation site
RNAPII	ribonucleic acid polymerase II
SF1	steroidogenic factor 1
StAR	steroidogenic acute regulatory protein
START	StAR-related lipid transfer domain
TSPO	18 kDa translocator protein
TSS	transcription start site
UTR	untranslated region
VDAC1	voltage-dependent anion channel

Introduction

StAR Perspectives: A History of the Discovery of StAR by Douglas M. Stocco

I would first of all like to point out that this narrative is being written from strictly a personal viewpoint. I cannot speak for what others were thinking during the past two to three decades with regards to the search for the putative regulator of acute steroidogenesis, so I will not try to do so. I came into the search for this regulator from a rather circuitous route and I will try to provide you with the events of how that took place in our laboratory. As such, I hope that you will understand that my position is certainly not to slight the accomplishments of others but merely to point out that I am not speaking for the entire field, but rather, only for myself and the members of our laboratory.

Some of the first observations on the production of steroids in response to trophic hormone treatment occurred in the laboratory of Oscar Hechter who studied the production of steroids in response to ACTH stimulation in perfused adrenal glands [1–4]. Following those early observations, James Ferguson's laboratory became engaged in similar studies [5, 6] and wrote in an article in *The Journal of Biological Chemistry* in 1963 "*of the several possible explanations for the observed effects of puromycin, the most provocative, but the most difficult to prove is the idea that a specific protein must be synthesized in order for the adrenal to increase steroid output.*" It was these words that seemed to sum up the problem at hand in understanding how the acute biosynthesis of steroids was regulated. Between Ferguson's early experiments and observations and those that followed by Garren [7–9]; Davis [10, 11]; Hall [12–15]; Brownie [16, 17]; Jefcoate [18–21]; Farese [22–26]; Boyd [27–31]; Koritz [32, 33]; Cooke [34–37]; and others, a generalized description of the characteristics of the protein predicted by Ferguson began to arise. It was thought that the putative protein was a trophic hormone-induced, rapidly synthesized, cycloheximide-sensitive protein that had a short half life and was thought to function by mediating the transfer of the substrate for all steroid hormones, cholesterol, from the outer to the inner mitochondrial membrane where it would be converted to pregnenolone by the cholesterol side-chain cleavage enzyme system that resided in the mitochondria.

Personally, I was not trained in the field of steroidogenesis nor was I trained in reproductive biology, but would soon have to learn something about both of these disciplines. I began working in the field of reproductive biology while engaging in collaborative studies with my colleague, Jim Hutson. Together, Jim and I were able to secure funding from the National Institutes of Health (NIH) and National Science Foundation (NSF), something that I had not been able to do on my own. I was interested in mitochondrial biogenesis but had difficulty in convincing the NIH of the virtues of studying how this fascinating organelle replicated itself in cells. In reading the literature on adrenal and gonadal steroidogenesis, I learned that a key step in the biosynthesis of steroids occurred in the mitochondria. In reading the works of those individuals named above, I thought that I might be able to parlay my keen interest in mitochondria into a new direction in which I might have better luck. During the course of my reading, I learned that new protein synthesis played a very important role in the regulation of the biosynthesis of steroids and that mitochondria were intimately involved in this regulation as well. I also learned that a number of investigators had sought this putative newly synthesized regulator, but that no definitive proof of its identity existed. Having run many 1-D electrophoresis gels and staining them with coomassie blue, I felt that finding a protein that might be present in trace amounts probably could not be done using this technology. An approach that would separate proteins to a greater extent and utilized a more sensitive detection method would probably be needed if the identity of this protein were to be uncovered. This approach had yielded positive results in studies by Nan Orme-Johnson and her colleagues who demonstrated the rapid synthesis of a series of phosphoproteins in response to ACTH stimulation of adrenal cells in culture by labeling the cells with ^{35}S-methionine and separating them on 2-D polyacrylamide gel electrophoresis (PAGE) [38–45]. This turned out to be the first demonstration of the protein that was later cloned in our laboratory, but I must sadly admit to not being aware of her elegant work at the beginning of my own work.

At about the time when I was planning to work on this project, I wondered if it would it be possible to use 2-D PAGE followed by sensitive silver staining to demonstrate the presence of proteins in stimulated steroidogenic cells that were not present in unstimulated cells, and thus give us a clue to the identity of the putative regulatory protein. We began these studies in my laboratory using primary cultures of rat Leydig cells that we isolated using commonly accepted methods. A great number of protein "spots" were detected on the gels and we attempted to search for differences between unstimulated and stimulated cells that would appear rapidly and be consistently reproducible. This, we were unable to do, despite many attempts. While we thought we had candidate proteins, our results were not consistent, and obtaining sufficient numbers of Leydig cells was proving to be time-consuming and costly.

In 1981, Mario Ascoli published a manuscript in which he described the isolation of several clonal lines of Leydig cells from the M5480P tumor [46]. One of these lines, the MA-10 cell line had functional hormone/chorionic gonadotropin (hCG) receptors and produced copious amounts of steroid in response to hCG, cholera toxin, and 8-Br-cAMP. This cell line seemed to be an ideal tool with which to continue our studies on the search for the regulator. Mario was most generous in sharing this cell line with us and we began to shift our efforts from primary cells to Leydig tumor cells. We found that the cells grew well in our hands and produced high levels of progesterone (due to a lack of cytochrome P450 17A1, CYP17A1) in response to stimulation. We also saw that they synthesized very low basal amounts of steroid in the absence of stimulation, another perceived advantage for us since the acute regulator may be present in small amounts without stimulation and in high amounts following stimulation. We reasoned that this might allow us to see differences between unstimulated and stimulated cells more clearly. We began these studies still utilizing 2-D PAGE and silver staining but the results we obtained remained equivocal and unsatisfactory, not allowing us to make any definitive statements concerning a potential acute regulator. We decided to shift from the use of silver staining to autoradiography as a detection method with the rationale that the putative regulator was newly synthesized and therefore may be radiolabeled to a higher degree than other proteins following stimulation. We tried labeling the cells first using a ^{3}H-amino acid mixture and then ^{14}C-leucine. Again, these approaches were not satisfactory. It was around that time that I became aware of the Orme-Johnson approach and we decided to utilize incubation in ^{35}S-methionine as a labeling procedure. It did not take too long to observe in stimulated MA-10 cells the proteins that we referred to as the 30 kDa proteins and which were, in all probability, identical to the proteins that Orme-Johnson had seen earlier in cultured adrenal cells. In the middle of these studies, I took a sabbatical leave in Holland in 1986 and joined Henk van der Molen's group at Erasmus University in Rotterdam. I did not work on the acute

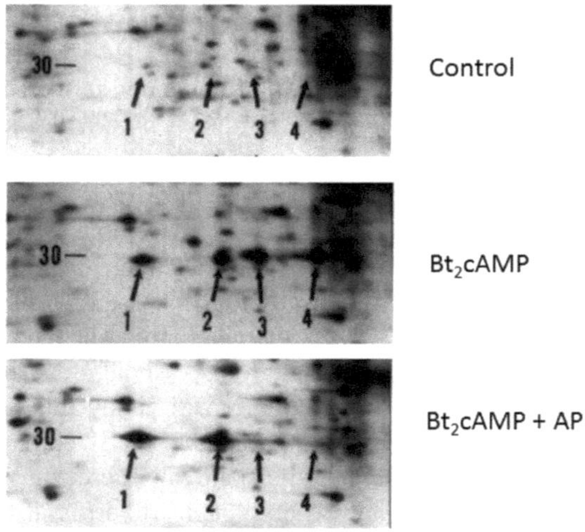

Fig. 2.1 Detection of the 30 kDa StAR protein by ^{35}S-methionine labeling and 2-D-PAGE. MA-10 mouse Leydig tumor cells were treated for 6 h with serum-free media containing ^{35}S-methionine in the absence (control) or presence of the cAMP analog dibutryl-cAMP (Bt_2cAMP, 1 mM). Mitochondria were isolated and the proteins separated by 2-D-SDS-PAGE. Shown are typical fluorograms of radiolabeled "spots on a gel" from this type of experiment. The arrows labeled 1–4 indicate the position of the StAR spots. The 30 kDa protein spots 1–4 appeared after cAMP treatment and correlated with steroid output (see text). Treatment of the isolated mitochondria with alkaline phosphatase (AP) resulted in loss of proteins 3 and 4 and recovery of radiolabel in protein spots 1 and 2. These data support that StAR is a phosphoprotein. Western blot analysis was used and confirmed that the protein spots 1–4 were StAR protein. *StAR* steroidogenic acute regulatory protein, *cAMP* cyclic adenosine-3′,5′-monophosphate

regulator while in Holland but did manage to learn a lot about male reproduction as the entire group in Rotterdam focused on this topic. I worked most closely with Focko Rommerts who had a great deal of experience in steroidogenesis and was quite familiar with the problem that I was trying to solve. We had many discussions concerning this topic and readily agreed that this was a difficult problem and that I would probably not be successful in solving it.

Upon returning to Texas, I began to work on the acute regulator once again. Our methodological approach was working well and we were readily able to reproduce the appearance of the 30 kDa proteins on 2-D gels and because of the purchase of a new computer-assisted imaging system, we were also able to accurately quantitate the proteins under many differing experimental conditions. We used this approach to do a series of experiments over the next several years to demonstrate that the 30 kDa proteins were induced by trophic hormone and cAMP analog, were rapidly synthesized in response to stimulation, were cycloheximide sensitive, were dose and time responsive to stimulation, and were localized to the mitochondria, the site of the regulated step [47–52] (Fig. 2.1). We also learned, at about the same time that the Orme-Johnson laboratory did, that the 30 kDa proteins were derived from a larger 37 kDa precursor much the same as many other mitochondrial proteins are derived, having an N-terminal targeting sequence that was removed during import into the mitochondria [39, 52]. All of the experiments that we performed were supportive of the idea that these proteins might be involved in the acute regulation of steroidogenesis, but as I was so abruptly reminded by a colleague upon giving a talk at the European Testis Workshop one year, "yes Doug, but they are just spots." Of course, we understood that correlations do not provide proof of anything and that we needed a cause and effect relationship between the 30 kDa mitochondrial proteins and steroid synthesis in order to convince anyone that these proteins were involved in the acute regulation of steroid biosynthesis. It was clear that running a hundred or a thousand more 2-D gels that supported our hypothesis would not provide the proof we needed, so we began to think about cloning the complementary deoxyribonucleic acid (cDNA) for this protein. I say this protein because at this time, we knew that while there were 4, 30 kDa "spots" on the gels, these spots were in fact modifications of the same protein and thus were one gene product. This would ostensibly make our task a bit easier, though still difficult.

I personally had no formal training in molecular biology and therefore the task of cloning this cDNA seemed daunting to me. I considered spending some time in Dallas where I was offered space in Evan Simpson's laboratory at the University of Texas Southwestern Medical Center, where I could work with my former student, Mike Kilgore, and learn the art of cloning and see if we could clone the cDNA for this protein. At that time, one of the most fortuitous and important things in my career happened. Barbara Clark had agreed to join my laboratory in 1992, just having finished her Ph.D. in Mike Waterman's laboratory at UT Southwestern in Dallas. Barbara was very well versed in all of the methodology that would be required to clone the cDNA for this protein. We talked about the project and she agreed to tackle it head on. It was an extremely difficult project as this work was performed prior to mass spectrometry and Matrix Assisted Laser Desorption Ionization Time-of-Flight (MALDI-TOF) and the fact that this protein represented approximately 0.2 % of the total cell protein and 0.7 % of the mitochondrial protein (as we later determined). Barbara diligently tried to purify the protein using fast protein liquid chromatography (FPLC). As we had no assay that we could utilize on the fractions collected, she attempted to "find" the protein by running mini 2-D gels on each of the eluted fractions, an unbelievable effort when you think about it. FPLC proved to be unsuccessful and Barbara then went through the arduous process of preparing mitochondrial mitoplasts, solubilizing the protein, running prep gels, cutting out the 28–32 kDa regions of the gels, extracting the proteins from these gels, concentrating the proteins, and running these concentrates on 2-D gels followed by silver staining to visualize the 30 kDa proteins. Then the proteins were carefully excised from the gels and retained until sufficient amounts of protein were obtained for microsequencing by the Harvard Microchemical Facility. The facility indicated to us that approximately 100 pM would be required for sequencing and approximately twice this amount was sent to them. Tryptic digestion of the 30 kDa protein, high-performance liquid chromatography (HPLC) separation of digested protein into separate peaks and sequencing of three of these peaks produced sequences of 19, 13, and 12 amino acids. With this information, degenerative nucleotides of 17–24 bases long were synthesized and used to amplify the 30 kDa cDNA from a cDNA library using polymerase chain reaction (PCR). This resulted in a 400 base pair product that in turn was used to probe the cDNA library once again. This probe resulted in the isolation of a 1,456 base pair full-length cDNA clone that when sequenced was found to represent a novel cDNA. The open reading frame of this cDNA was found to be 852 base pairs in length and coded for a protein that was 284 amino acids in length that also proved to be a novel sequence. A number of steps were taken to assure that the correct cDNA had been cloned with perhaps the most definitive proof being that the open reading frame contained the exact same sequences that were provided to us by the Harvard Facility. While this all sounds very straight forward, I can assure you it was anything but that, as we encountered a number of starts and stops all along the way during these efforts, just as all laboratories do.

We were now at last in a position to determine if a cause and effect situation existed for this protein. Barbara cloned the cDNA for the 37 kDa precursor protein into a pCMV expression vector and transfected this vector into unstimulated MA-10 cells and compared steroid synthesis to empty pCMV-transfected cells. Expression of the 37 kDa/30 kDa protein resulted in a twofold to fourfold increase in progesterone production in the absence of hormone stimulation, and thus provided the cause and effect observation that was required to demonstrate the function of this protein. We strongly felt that this was the evidence that we needed to convince the scientific community that this protein functioned as the acute regulator of steroid hormone biosynthesis that had long been sought. We named the protein StAR, for Steroidogenic Acute Regulatory protein. During the course of these experiments, other individuals in my laboratory added significant observations to this story. Steven King performed in vitro transcription and translation of the 37 kDa cDNA and showed that this resulted in the synthesis of all 4 forms of the 30 kDa protein, a strong supporting piece of information [53]. Jeff Wells, a technician in our laboratory, worked tirelessly in growing and harvesting MA-10 cells that were needed for the isolation of the 30 kDa proteins. This was a team effort that culminated with the publication describing the purification, cloning, and expression of the StAR protein in the November 1994 issue of *The Journal of Biological Chemistry* [53].

Earlier in 1994, I met Jerry Strauss from the University of Pennsylvania, at a meeting in the UK. I was discussing some of our results with Jerry and eventually the discussion came around to the enigma of searching for the cause of the disease, lipoid congenital adrenal hyperplasia (lipoid CAH). Lipoid CAH results from an almost complete inability of the afflicted patients to synthesize steroid hormones. Lipoid CAH was originally thought to be a result of a mutation in the CYP11A1 gene, but this proved to be incorrect once the human gene was cloned and later sequenced from the tissue of patients having this disease [54]. Similarly, other protein candidates were discounted as the cause of lipoid CAH. We felt that malfunction of the 30 kDa protein that we were working on could be a candidate for causing lipoid CAH. We then sent the 400 bp cDNA fragment to Jerry and his laboratory was able to clone the human cDNA counterpart of the StAR gene. The information that resulted from cloning and sequencing the human StAR cDNA was sent to another colleague, Walter Miller at the University of California, San Francisco. Walter had access to human tissue (testicular tissue) from patients who had been diagnosed with lipoid CAH. Using the sequence information he received from Jerry, Walter's laboratory was able to clone the StAR cDNA from lipoid CAH tissue. The first four clones that Walter's lab isolated and sequenced from lipoid CAH patients had mutations in the StAR cDNA. Cloning these mutated cDNAs into expression vectors and expressing them in COS-1 cells, transfected with the cholesterol side-chain cleavage enzyme system, indicated that these mutated StAR cDNAs could not support steroid synthesis while control StAR cDNAs could. In essence, nature had provided a StAR knockout and its phenotype proved that StAR was an essential element for acute steroid biosynthesis. These elegant and important results support the role of StAR and also provided unequivocal evidence for the cause of lipoid CAH. These results were published in *Science* in the spring of 1995 [55]. The human StAR mutation findings were corroborated in 1997 with the Proceedings of the National Academy of Sciences USA (PNAS) publication of the phenotype of the mouse StAR knockout by Keith Parker's laboratory [56]. The mouse phenotype essentially mirrored the human phenotype and thus, a rodent lipoid CAH model became available for further study. Since that time, less severe mutations in the StAR gene as well as mutations in the CYP11A1 gene have been identified in patients with lipoid CAH [55, 57–59].

These are the best recollections of the history of the discovery of StAR that I have as they pertain to my own laboratory. In a narrative such as this, it is impossible to chronicle each of the individual contributions made to this story and I apologize for leaving out the names of individuals who have so contributed. Individual contributions can, of course, be gleaned from the publications that have chronicled this series of investigations. As of this writing, more than 1730 manuscripts have been published that were identified using "steroidogenic acute regulatory" as a probe on PUBMED. Last, but certainly not the least, I would like to acknowledge the support of the National Institute of Child Health and Human Development (NICHD) of the NIH and the Robert A. Welch Foundation for making this work possible.

StAR (STARD1) Overview

Over a 20-year span (1994–2014), the biochemical and genetic data on StAR has confirmed its essential role in cholesterol transfer into mitochondria and in controlling steroidogenesis (reviewed in [60]). Of particular importance was the finding that mutations in the *hSTARD1* gene are the most common basis for lipoid CAH, as discussed above [55, 61]. The mechanism of cholesterol transport by StAR and lipoid CAH is the topic of Chap. 4 so we will only briefly summarize StAR protein function. StAR is a nuclear gene that encodes a protein that is synthesized in the cytoplasm as a precursor protein with an amino-terminal domain characteristic of a mitochondrial targeting sequence; e.g., a predicted amphipathic helix commonly found in matrix localized proteins [53]. Mitochondrial import and processing of the newly synthesized 37 kDa precursor StAR produces a 32 kDa intermediate product and a mature 30 kDa matrix localized form (reviewed in [60]). The current model for StAR function is that StAR is active on the cytosolic side of the outer mitochondrial membrane and that import and processing "inactivates" StAR. The inactivation is a consequence of localization of StAR to the interior of the mitochondria rather than a loss of cholesterol binding and function. Indeed, key studies showed that in the absence of the N-terminal mitochondrial targeting sequence, StAR is composed solely of a START domain that does not get processed yet is capable of binding cholesterol and facilitating cholesterol transport across mitochondrial membranes [62]. Therefore, the import and processing of StAR by mitochondria is proposed to be an "off" switch for cholesterol transfer and thus steroidogenesis. However, studies from StAR knockout mice and StAR transgenic animals implicate that in vivo the targeting of StAR to mitochondria may be important for efficient cholesterol transfer and steroid production. StAR knockout mice accumulate significant amounts of cholesterol in the adrenals and gonads and the animals die shortly after birth due to the absence of adrenal hormones [63]. Lipid accumulation was attenuated and adrenal and gonadal steroidogenesis were restored in StAR knockout mice that expressed a StAR transgene. On the other hand, when an amino terminal truncated StAR transgene was expressed, the mice retained lipid accumulation in the adrenal and gonads and had partially restored steroidogenesis in a tissue- and gender-specific manner [64]. The StAR transgenic animal studies support that StAR is capable of functioning without being targeted to the mitochondria, but highlight the importance of correct and efficient subcellular localization of the protein for full function in vivo.

Since continual synthesis of 37 kDa precursor StAR is required to maintain the active cytosolic form of StAR associated with mitochondria, the mechanisms that control StAR protein expression control steroidogenesis. Both transcriptional and posttranscriptional mechanisms regulate StAR gene expression, and this chapter provides an overview of the mechanisms controlling StAR expression in the gonads and adrenal. First, a general overview for transcription initiation and hormonal regulation of steroidogenesis is provided to set the context for StAR regulation.

Basic Mechanisms for Transcription Initiation

A gene that is transcribed by RNA Polymerase II (RNAPII) has distinctive functional elements that are required to initiate and enhance transcription. The promoter region of a gene contains sequence-specific cis-acting DNA elements, or regulatory elements, that are recognized and bound by trans-acting DNA-binding proteins (transcription factors) required to initiate transcription. Studies that characterize gene promoters provide the location for these cis-acting regulatory elements and, by convention, the position of these elements are designated relative to the transcription start site (TSS). The TSS is designated as the first nucleotide transcribed into pre-mRNA and is denoted numerically by +1. The core promoter is defined as the minimal region required for RNAPII to initiate transcription. There are two major classes of core promoter sequences for RNAPII binding; an AT-rich DNA sequence containing a consensus TATAAAA element called the TATA box is found in approximately 25 % of mammalian genes, and CG-rich regions referred to as a TATA-less promoter found in the remaining 75 % of the genes. In general, TATA-box elements are found in highly regulated, cell-specific expressed genes while CG-rich regions serve as promoters for many housekeeping or commonly expressed genes. Recruitment of RNAPII to the core promoter and proper transcription initiation is controlled by the assembly of the general transcription complex that together with RNAPII forms the transcription preinitiation complex (PIC). The assembly of the PIC is enhanced by activator and coactivator proteins that bind to regulatory elements within the promoter.

Activator proteins are transcription factors that bind their respective sequence-specific DNA regulatory elements and function in either a constitutive manner (always bound to DNA) to maintain steady-state levels of gene expression or in a developmental or regulated manner to activate gene transcription in response to external stimuli such as hormonal signaling. The sequence-specific DNA regulatory elements typically contain a twofold symmetry and the activator proteins bind as homo- or heterodimers. Common DNA and protein interacting motifs identified within activator proteins include the helix-turn-helix (HTH), basic helix-loop-helix (bHLH), zinc finger (Zn finger), and basic leucine zipper (bZIP) motifs. bZIP proteins have a conserved region aligned by the spacing of leucine resides every 7th amino acid that folds into an α-helix containing a hydrophobic face with evenly spaced leucines. The interdigitation of the leucines between these α-helices forms the leucine zipper and promotes protein dimerization. A basic region is adjacent to each helix and binds the half-site of the response element; therefore, dimerization is necessary for function (DNA binding). bZIP motifs are commonly found in activators that are regulated in response to acute cellular signaling. For example, the c-FOS and c-JUN proteins of the AP-1 transcription family are phosphorylated and activated by protein kinase C (PKC) signaling and the cAMP responsive element-binding protein, CREB, is phosphorylated and activated by the protein kinase A (PKA) signaling pathway. Another common DNA-binding motif is the zinc finger; a structure formed by the coordination of Zn ions by specifically spaced cysteine (Cys4) or cysteine and histidine (Cys2His2) residues that generates adjacent finger-like structures that bind the major groove of DNA and promote protein dimerization. The Zn finger Cys_2Cys_2 motif is shared by members of the steroid hormone/nuclear receptor superfamily where ligand binding enhances homo- or heterodimer formation and DNA binding.

Coactivator and corepressor proteins do not bind DNA directly but are recruited to promoters via protein interactions with the DNA-bound activator proteins (reviewed in [65–67]). Conserved protein interacting motifs found on activating proteins include the LXXLL motif for nuclear receptor-coactivator interactions, and kinase interacting domain (KID) for CREB/activated transcription factor (ATF) family members and coactivators [67, 68]. Protein-protein interactions can be dependent upon posttranslational modification(s), such as phosphorylation of CREB on Ser133 by PKA or Ca^{2+} signaling, which results in increased association with the coactivator CREB-binding protein (CBP/p300 [68–71]). Coactivators classically have intrinsic histone acetyltransferase activity (HAT) and acetylate core histone proteins (H2A, H2B, H3, and H4) leading to open chromatin structure for active transcription. Members of the HAT family include CBP/p300, P/CAF (p300/CBP Associated Factor), and the p160 family that includes SRC-1(NCoA-1), SRC-2 (NCoA-2), and SRC-3. Additional histone modifying enzymes have been shown to contribute to coactivator function [72].

Conversely, a gene may be actively silenced by the constitutive presence of a corepressor complex on the promoter. Corepressor proteins, like coactivators, are associated with DNA bound activator proteins via protein-protein interactions thereby localizing the corepressor to specific promoter regions of the gene [67]. The repressor protein then recruits HDAC proteins, enzymes with intrinsic histone deacetylase activity that help maintain histones in the deacetylated state and promote gene silencing. mSin3 and nucleosome remodeling and histone deacetylation (NuRD) are two common repressors utilized by multiple activator proteins to recruit HDAC1 to the promoter. Nuclear receptors also can utilize the corepressors silencing mediator of retinoid and thyroid receptors (SMRT) and nuclear receptor corepressor (NCoR) to either recruit HDACs directly or indirectly via Sin3A complex formation [73, 74].

Hormonal Regulation of Steroidogenesis and StAR Function

The hypothalamic-pituitary-gonadal axis is the major hormonal signaling cascade that controls sex hormone output by the testis and the ovary while the hypothalamic-pituitary-adrenal axis is the major hormonal signaling cascade that controls glucocorticoid synthesis by the adrenal gland (Fig. 2.2a). In brief, the hypothalamus secretes gonadotropin-releasing hormone (GnRH) or corticotropin-releasing hormone (CRH) that then stimulate the gonadotropes or corticotrophes of the anterior pituitary, respectively, to synthesize and secrete the gonadotropins leutinizing hormone [75] and follicle stimulating hormone (FSH) or adrenocorticotropin hormone (ACTH). LH stimulates testosterone synthesis by testicular Leydig cells and androgen production by ovarian theca cells while FSH stimulates estrogen production by ovarian granulosa cells. ACTH stimulates glucocorticoid synthesis by the adrenal gland. Mechanistically, ACTH, LH, and FSH bind to their respective 7-transmembrane G-protein-coupled receptors in their target tissues and activate signal transduction pathways leading to increased expression and/or activation of transcriptional activator proteins and subsequent increase in StAR and other genes within the steroid hormone biosynthetic pathway (see Fig. 2.2b and legend for more detailed description). The cross-talk between these pathways leads to maximal increase in steroid output. The cAMP-PKA signal transduction pathway is central to

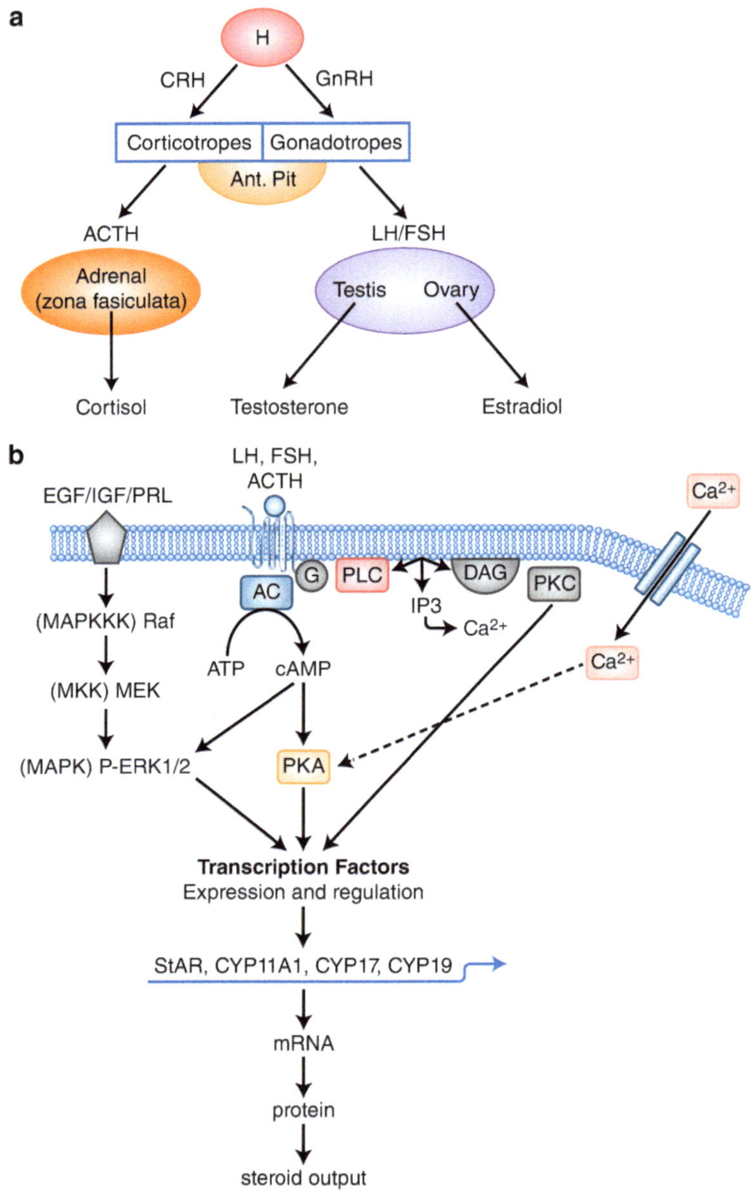

Fig. 2.2 Mechanisms for tropic hormone regulation of steroidogenesis. **a** The hypothalamic-pituitary-adrenal/gonadal axis controlling steroid hormone production. The pathway is described in the text. **b** Mechanisms for increased StAR and steroidogenic gene expression following tropic hormone stimulation. Shown is a model of the major signal transduction pathways activated by tropic hormone (LH, FSH, ACTH) stimulation of their respective G-protein-coupled receptors. Activated second messenger pathways include adenylyl cyclase (AC)-mediated increase in cAMP levels and the subsequent activation of protein kinase A (PKA), phospholipase C (PLC)-mediated increase in diacylglycerol (DAG) and inositol 1, 4, 5 trisphosphate (IP3). DAG then activates protein kinase C (PKC) while IP3 increases intracellular calcium levels that also promote PKC activation. Epidermal growth factor (EGF), insulin-like growth factor (IGF), and prolactin (PRL),

hormone-dependent activation of gonadal sex hormone and adrenal glucocorticoid and androgen synthesis. In addition, aldosterone, a critical steroid hormone that regulates salt balance and blood pressure, is produced by the adrenal gland. Major regulators of aldosterone synthesis are angiotensin II (Ang II) and K^+. Ang II is a peptide hormone that binds to its respective G protein-coupled cell surface receptor and activates the protein kinase C (PKC) and calcium signaling pathways while K^+-mediated membrane depolarization activates voltage-gated calcium channels and subsequent calcium-PKC- signaling pathways (reviewed in [76]). Similar to cAMP-PKA signaling, Ang II and K^+ lead to transcription factor phosphorylation and increased target gene expression, including StAR, resulting in increased aldosterone synthesis [77–80]. This chapter focuses on the cAMP-PKA-dependent mechanisms that activate StAR gene expression. The unifying theme for StAR gene regulation is that hormone-dependent activation of the signaling pathways that regulate transcription factor activation and recruitment to the StAR promoter functions in a cell-type-dependent manner resulting in cell-specific mechanisms that control StAR expression.

PKA-dependent phosphorylation of StAR is a key mechanism controlling StAR function. StAR was originally observed to be a phosphoprotein, a modification that was established by the early observations of four protein spots after 2-D gel electrophoresis; two of which were sensitive to alkaline phosphatase treatment (Fig. 2.1). Two consensus PKA phosphorylation sites at Ser56/57 and Ser194/195, in murine and human StAR, respectively, are *bona fide* PKA targets [81, 82]. StAR protein harboring a Ser194/95Ala mutation fails to mediate cholesterol transport into mitochondria, although the cholesterol-binding property is not changed due to phosphorylation [81–83]. In addition, PKC-dependent synthesis of wild-type StAR without subsequent PKA-dependent phosphorylation does not promote steroid production, supporting a role for posttranslational modification of StAR for function [83]. While the importance for StAR phosphorylation is established, the mechanistic link between StAR phosphorylation and function remains to be determined.

---

activate the Ras/Raf and mitogen-activated protein kinases (MAPK/ERK) pathways, which have been shown to modulate the cAMP-PKA-dependent control of StAR expression. The activation of these protein kinases results in the increased expression and/or phosphorylation of transcription factors that control StAR and steroidogenic enzymes within the steroidogenesis pathway, leading to an increase in steroid output. The potential cross-talk between these pathways through cAMP-mediated activation MAPK or calcium signaling modulating PKA illustrates the potential for multi signals converging on activation of key transcription factors that contribute to increased gene expression. The cAMP-PKA pathway is the key pathway regulating StAR expression and steroidogenesis in steroidogenic tissues. This may be attributed, in part, to the PKA-dependent phosphorylation of StAR that is required for function. The model pathway in **b** is adapted from [114]. CYP11A1, cytochrome P450 side chain cleavage enzyme; CYP17, cytochrome P450 17a-hydroxylase, 17, 20 lyase; CYP19, cytochrome P450 aromatase are used as examples of regulated genes within the steroid hormone biosynthetic pathway *H* hypothalamus, *CRH* corticotrophin-releasing hormone, *GnRH* gonadotropin-releasing hormone, *Ant. Pit.* anterior pituitary, *ACTH* adrenocorticotropin hormone, *LH* luteinizing hormone, *FSH* follicle stimulating hormone, *StAR* steroidogenic acute regulatory protein, *cAMP* cyclic adenosine-3′,5′-monophosphate, *DAG* diacylglycerol

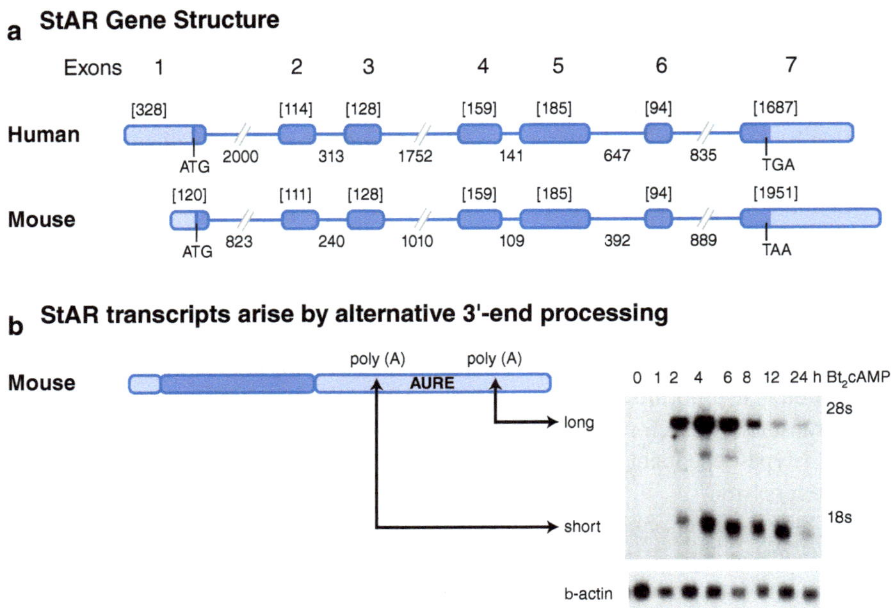

Fig. 2.3 StAR gene and primary transcript structure. **a** Schematic representation of the human and mouse StAR gene structure. The sequence information obtained from human gene ID 6770, NG_011827 and mouse gene ID 20845, NM_011485 were used to cartoon the gene structures for human and mouse StAR, respectively. Exons are depicted by dark boxes with the 5′ and 3′ untranslated regions shown in lighter color and the size of the exons are given above each box. Introns are depicted by a line with the size shown below the line. The start (ATG) and stop (TGA/TAA) codons are indicated in the schematic. Exon 2 of human StAR contains an additional 3 base pairs relative to the mouse sequence. **b** Alternative 3′-end processing results in two major StAR transcripts. Shown is a schematic of the mouse StAR primary transcript. The coding region is depicted by the dark box with the 5′ and 3′ UTR in lighter shade. Two major polyadenylation signals (poly(A)) within the 3′-UTR of mouse StAR that promote differential processing leading to a long (~3.5 kb) and short (~1.6 kb) transcript in rodent cell lines. The two major transcripts are shown by Northern blot analysis of StAR mRNA in MA-10 mouse Leydig tumor cells after treatment with the cAMP analog Bt_2 cAMP for the indicated time. The presence of destabilizing AUREs (AU-rich elements) in the longer transcript are proposed to contribute to the shorter half-life of the longer transcript [173]. Figure for Northern blot data is adapted from data presented in [84] *StAR* steroidogenic acute regulatory protein, *mRNA* messenger ribonucleic acid, *cAMP* cyclic adenosine-3′,5′-monophosphate, *AU* adenosine-uridine

StAR Gene Structure

Shortly after we cloned the mouse StAR cDNA, we characterized the mouse gene [84]. The gene is comprised of 7 exons and 6 introns that span a relatively short distance, 6.25 kb, located on chromosome 8 [84]. The cDNA contains an open reading frame of 852 bases encoding a protein of 284 amino acids with predicted molecular weight of 31.6 kDa [53]. The human StAR gene shares the same general structure with 7 exons-6 introns spanning ~8 kb that maps to chromosome 8p11.23 [85] (Fig. 2.3). The human cDNA encodes a protein of 285 amino acids due to an additional 3 nucleotides in exon 2 that results in an insertion of a serine at position 47. Many of the mammalian StAR proteins are 285 amino acids in length, with the mouse and rat being the apparent exceptions. The additional amino acid has no apparent functional advantage since deletion of the amino terminal 65 amino acids from StAR does not decrease activity for cholesterol transfer.

The core promoter of StAR contains a TATA-box, common for highly regulated genes. To localize the cAMP-responsive region of StAR promoters, early studies used reporter gene assays and tested the activity of a series of 5′-deletion constructs beginning with >3 kb of the 5′-flanking sequence upstream of the transcription start site (TSS). A region within 250 bp of the TSS was shown to be sufficient for maximal promoter activity of mouse, human, rat, and bovine StAR genes in the adrenal and gonads. Alignment of StAR promoter regions from multiple species revealed highly conserved sequence-specific DNA regulatory elements within the first 150 bp region immediately upstream of the TSS (Fig. 2.4a). This proximal promoter region of StAR has been most extensively studied utilizing mouse and human cell lines. These studies demonstrated that the conserved regulatory elements bind their cognate transcription factors in a cell-type- and species-specific manner. The major transcription factors found to activate StAR transcription are steroidogenic factor-1 (SF-1 also referred to as adrenal 4-binding protein and encoded by NR5A1), NUR77 (NR4A1), CCAAT/enhancer-binding protein beta (C/EBPβ), GATA-4, activator protein-1 (AP-1) family members, and CREB/CREB-modulator (CREM) family members (Fig. 2.4a).

Zinc-Finger Transcription Factors Involved in StAR Gene Expression

SF-1 is an orphan member of the nuclear receptor superfamily that binds as a monomer to the consensus DNA sequence, 5′- AGGTCA-3′; a sequence that was originally found to be common in promoters of several steroid hydroxylase genes [86–89]. SF-1 is highly expressed in steroidogenic cell types where it functions to help control the tissue-specific expression of genes involved in the steroid hormone biosynthesis pathway [90, 91]. Additionally, gene knockout studies in mice showed that SF-1 is critical for development and differentiation of the endocrine and reproductive systems, demonstrating it has multiple functions [92]. SF-1 function has been linked to both basal and cAMP-PKA-dependent gene expression, with promoter- and cell-specific mechanisms contributing to SF-1-dependent target gene expression. Mutational analysis of a non-consensus SF-1 element located directly upstream of the CRE/AP-1 element in StAR proximal promoter verified SF-1 plays a major functional role in StAR expression (Fig. 2.4). SF-1 binds the RNA coactivator SRA (steroid receptor activator RNA) and SRA is important for SF-1-dependent gene activation in the adrenal and gonads [93]. While the cAMP-PKA-dependent mechanisms controlling SF-1 function are not fully elucidated, current working

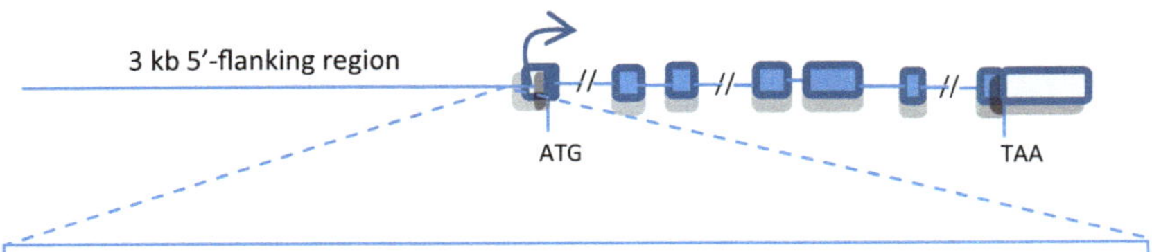


```
                                   SF-1                                     C/EBPβ
ACCCTCCCCCG--CCCCAAGCTCCCCACAAACGGCCAAAGCAGCAGTGTGAGGCAATCGC   human
                   TCCCTCCCACCTTGGCCAGCACTGCAGGATGAGGCAATCAT   mouse
                            CCTTGGTCAGCACTGCAGTATGAGGCAATCAT   rat
ACCCTCCCTCTCCACCCCAGCTCCCCGCAAATGGCCAAAGACTCCTGGTGAGGCAATCTC   bovine
--CCTCCCCCC--CC-CCATCCCCCGCCAAATGCCCAAAGGCTGCTTGTGAGGCAATCCC   pig
ACCCTCCCCCG--CCCCAAGCTCCCCACAAATGGCCAAAGCAGCAGTGTGAGGCAATCGC   chimp
  ******                   ..: * ***...     .  .**********

  SF-1/NUR77                  CREB/CREM/AP-1        GATA
TCTATCCTTGACCCCTTCCTTTGCACAG-TGAGTGATGGCGTTTTTATCTCCTGATGATG   human
TCCATCCTTGACCCTC-----TGCAC-AATGACTGATGACTTTTTTATCTCAA-GTGATG   mouse
TCCATCCTTGACCCTC-----TGCA-CAGTGACTGATGGCTTTTTTATCTCAA-GTGATG   rat
TCTATCCTTGGTGCCTTCCTTTCCAC-GCTGAGTGATGGCTTTTTTATCTCCCGATGATG   bovine
TCTATCCTTGGCCCCTTCCTTTGCAAGGTTGAGTGATGGTTTTTTTATCTCCAGATGATG   pig
TCTATCCTTGACCCCTTCCTTTGCACAG-TGAGTGATGGCGTTTTTATCTCCTGATGATG   chimp
** *******.  *        * **. . *** *****.  **********. . *****

       SF-1
ATGCACAGCCTTCAGCGGGGGACATTTAA -19                              human
ATGCACAGCCTTCCACGGGAAGCATTTAA -22                              mouse
ATGCACAGCCTTCCGCTGGAAGCATTTAA -25                              rat
AAACACAGCCTTCAGCTGGAGGTATTTAA                                  bovine
AAACACAACCTTCAGCTGGAGGCATTTAA                                  pig
ATGCACAGCCTTCAGCGGGGGGCATTTAA                                  chimp
*:.****.*****..* **... ******
```

a

```
     SF-1                     CREB/CREM/AP-1       GATA
TCTATCCTTGACCC ^ GCACAGTGAGTGATGGCGTTTTTATCTCCTGATGATG   human
TCCATCCTTGACCCTCTGCACAATGACTGATGACTTTTTTATCTCAA GTGATG   mouse
```

b

Fig. 2.4 Multiple Sequence Alignment of the StAR Proximal Promoter. **a** Shown is a ClustalW alignment of human, mouse, rat, bovine, pig, and chimpanzee StAR promoter sequences between the TATA box (TTTAA) and the distal SF-1 element located at -145 bp in the mouse promoter sequence. The binding sites for SF-1, C/EBPβ, CREB/CREM/AP-1 are shown in *light blue* and labeled with the names of the respective transcription factor(s). **b** Comparison of the cAMP-responsive region of the human and mouse StAR proximal promoter. Shown is a short 53 bp region that encompasses non-consensus SF-1 and CRE/AP-1 elements and a consensus GATA element. Each of these DNA elements has been shown to bind their respective transcription factors and increase StAR transcription. The divergence between the mouse and human sequence within the critical elements and in the flanking DNA sequences may contribute, in part, to promoter (species)-specific differences in transcription factor binding. *StAR* steroidogenic acute regulatory protein, *C/EBP* CCAAT-enhancer-binding protein—beta, *CREB* cyclic-AMP responsive element-binding protein, *CREM* CREB modulator protein, *AP-1* activator protein 1, *SF-1* steroidogenic factor-1, *cAMP* cyclic adenosine-3′,5′-monophosphate, *DNA* complementary deoxyribonucleic acid

models include indirect mechanisms by which cAMP-PKA signaling promotes production of SF-1 ligands and SF-1 cofactors to enhance SF-1 function [94].

NUR77, encoded by the NR4A1 gene and also referred to as nerve growth factor induced-B (NGFI-B) is an orphan nuclear receptor that binds as a monomer to the cis-acting consensus sequence 5′-AAAGGTCA-3′ called the NGFI-B response element (NBRE). As the name implies, NUR77, and related family members NOR1 and NURR1, were characterized for their function in the nervous system as part of the immediate early response to stimuli such as growth factors and membrane depolarization [95]. In steroid producing tissues NUR77 expression is increased by hormonal stimulation and NUR77 has been shown to regulate steroidogenic gene expression [96–102]. The NBRE shares sequence identity with the SF-1 element and a NBRE overlaps the SF-1 element in the mouse STAR promoter (Fig. 2.4a). Calcium signaling contributes to the cAMP-dependent increase in steroidogenesis and a calcium-dependent increase in NUR77 expression leads to increased binding to the SF-1/NBRE element in the mouse StAR promoter [102–104] (Fig. 2.4a). However, NUR77 does not bind this region in the human StAR promoter [105, 106], likely due to flanking sequence variation at the SF-1/NBRE element between the mouse and human promoters (Fig. 2.4a). The inability for NUR77 to bind to the human StAR promoter, and potentially compensate for SF-1 function, may help explain the dependence on SF-1 for StAR promoter activity in human cell lines [78, 107–110].

The GATA family of zinc-finger DNA-binding proteins is named for the cis-acting DNA core consensus sequence, 5′- GATA-3′, that they bind. This family is composed of six members, GATA-1–GATA-6, with GATA-4 associated with gene transcription and development in the heart. However, GATA-4 is expressed in steroidogenic tissues as well as in the heart, lung, liver, and small intestine and has been shown to play a broader role in gonadal development and tissue-specific gene expression [111–113]. GATA proteins function with FOG (friend of GATA) and p300/CBP to activate target gene expression. StAR and several steroid hydroxylase gene promoters contain GATA elements and GATA-4 has been shown to regulate these genes in the testis and ovary.

bZIP Transcription Factors Involved in cAMP-Dependent StAR Gene Expression

Comprehensive reviews of the role of bZIP proteins in StAR gene regulation summarize the studies that identified the roles for members of this family of transcription factors that bind to a non-consensus CRE/AP-1 element in the StAR promoter and control transcriptional activation or repression [114, 115]. Therefore, only a broad overview highlighting the current models for StAR gene regulation is provided herein.

The *CREB, CREM, and ATF-1* genes encode a family of transcription factors that bind to cAMP-responsive elements [116] that have the consensus sequence

5′-TGACGTCA-3′. CREB/CREM/ATF-1 family members can form homo- or heterodimers through the bZIP domain and can functionally compensate for each other in activating target gene expression in response to cAMP-PKA signaling [117–120]. CREB is known generally as a transcriptional activator while multiple CREM isoforms, generated via alternative splicing or use of alternative promoters of the CREM gene, both transcriptional activators (CREMτ, τ1, and τ2) or repressors (CREMα, β and γ) and ICER (inducible cAMP early repressor). CREMτ is the major transcriptional activator isoform in the testis, with strong expression in germ cells where it controls expression of genes critical for spermatogenesis [121]. In mouse Leydig cell cultures, both CREB and CREM are phosphorylated in response to PKA activation and the phosphoproteins bind to the StAR CRE/AP-1 element and contribute to the cAMP-dependent increase in StAR transcription [122, 123]. CREB is not expressed in the adrenal and in this tissue, CREM isoforms bind to the CRE/AP-1 element and activate StAR transcription [78, 110].

AP-1 is a transcriptional complex composed of Fos and Jun proteins. Members of the Fos family (c-Fos, FosB, Fra-1, and Fra-2) are found only as a heterodimer partner with members of the Jun (c-Jun, JunB, and JunD) proteins and with some members of the CREB/ATF family, while Jun members form either homodimers or heterodimers [124–126]. AP-1 family members are phosphorylated in response to PKA or PKC activation and also bind to the CRE/AP-1 element [122, 127]. c-Jun functions as a potent trans-activator of *StAR* transcription in Leydig cells, possibly through the formation of heterodimers with CREB family members [127], while c-Jun/c-Fos heterodimers can both activate and repress StAR [128–131].

The C/EBP family members bind to CCAAT box DNA elements with the consensus sequence A/GTTGCGC/TAAC/T [132]. The C/EBP family is best known for regulating genes that control cell differentiation and are differentially expressed in a tissue-specific manner [133–135]. In steroidogenic cells, C/EBPα and C/EBPβ isoforms are detected and C/EBPβ expression is increased by cAMP-mediated mechanisms in mouse testicular Leydig and ovarian granulosa cells [136–138]. The functional C/EBP element in the StAR promoter is highly conserved and C/EBPβ in MA-10 Leydig cell, rodent granulosa-luteal cell, and human granulosa-luteal cell nuclear extracts was shown to bind to the mouse and human StAR proximal promoter [139–141]. Protein-protein interactions between C/EBPβ and SF-1 and C/EBPβ and GATA-4 have been proposed to contribute to Leydig and granulosa cell StAR gene expression, respectively.

Model for cAMP-PKA-Dependent Activation of STAR Transcription

A cAMP responsive region, referred to as the CAN or CRE/AP-1 site, is a non-consensus CRE with overlapping specificity for members of the CREB/CREM and AP-1 protein families (Fig. 2.4b). This region is the most complex of the StAR proximal promoter due to the complex pattern of transcription factor binding that occurs in a species- and tissue-specific manner. Potential heterodimer formation between CREB, AP-1, and C/EBP bZIP proteins provide a level of control to integrate gene expression through one common element yet distinct trans-activating proteins.

One unifying theme identified by studies on StAR gene regulation from multiple laboratories is full promoter activity and a robust cAMP-dependent response involves SF-1 and GATA-4 binding to their respective elements in the StAR promoter (Figs. 2.4b and 2.5). Mutation of either the SF-1 or GATA-4 site does not diminish cAMP induction of StAR promoter activity, yet the loss of both of these factors attenuates overall promoter activity. The SF-1 and GATA-4 elements flank the critical CRE/AP-1 element, and protein-protein interactions between SF-1-GATA-4, SF-1-CREB, and C/EBPβ-GATA-4 have been reported and these interactions may help stabilize CREB/CREM/AP-1 factor binding at the CRE/AP-1 element. However, the fine-tuning of the hormone-dependent response is controlled by factor binding to the CRE/AP-1 element. Given that several combinations of bZIP proteins bind to this element, it is likely that the relative cellular levels of CREB/CREM/AP-1 proteins and preferential activation (phosphorylation) of these factors by hormone-dependent activation of the PKA and/or PKC signaling pathways regulate bZIP protein heterodimer partner formation controlling StAR gene expression in a cell type dependent manner.

The temporal pattern for transcription factor recruitment is nicely demonstrated by chromatin immunoprecipitation (ChIP) assays. The data support that upon 8-Br-cAMP treatment of MA-10 mouse Leydig tumor cells, the immediate response is recruitment of SF-1, GATA-4, CREB/CREM/AP-1, and CBP to the mouse StAR promoter followed by histone H3 acetylation and increased StAR mRNA levels. CREB, AP-1, and GATA-4 phosphorylation increases immediately following cAMP treatment of MA-10 cells with no change in total protein expression [142], indicating that phosphorylation of these factors promotes their recruitment to the StAR promoter and facilitates increased coactivator interactions [122, 140] (Fig. 2.5).

One model for cAMP-PKA stimulated StAR transcription that integrates the above studies is tropic hormone stimulation results in PKA-dependent phosphorylation of CREB/CREM and/or c-Fos/c-Jun (AP-1) and increases recruitment of these transcription factors to the CRE/AP-1 site in the StAR proximal promoter. SF-1 and GATA-4 bind to their respective elements that flank the CRE/AP-1 site. PKA-dependent phosphorylation of GATA-4 enhances both recruitment of this factor to the StAR promoter and protein-protein interactions. Protein-protein interactions between GATA-4 and SF-1, AP-1 and/or CREB and possible CREB- SF-1 interactions help stabilize the transcription complex and enhance recruitment of the coactivator CBP and RNAPII to the StAR promoter and increase transcription. SF-1-GATA-4 binding appear to be a common mechanism for mouse and human StAR gene activation in multiple cell types while the homo- or heterodimer partners (CREB/CREM or AP-1 or CREB-cJun) binding to the CRE/AP-1 element is dependent upon the cell type and stimulus.

An example of integrating StAR gene expression through one common element, yet distinct trans-activating proteins can be found in the follicular to luteal phase transition in the rodent ovary. During this transition, regulation of steroid

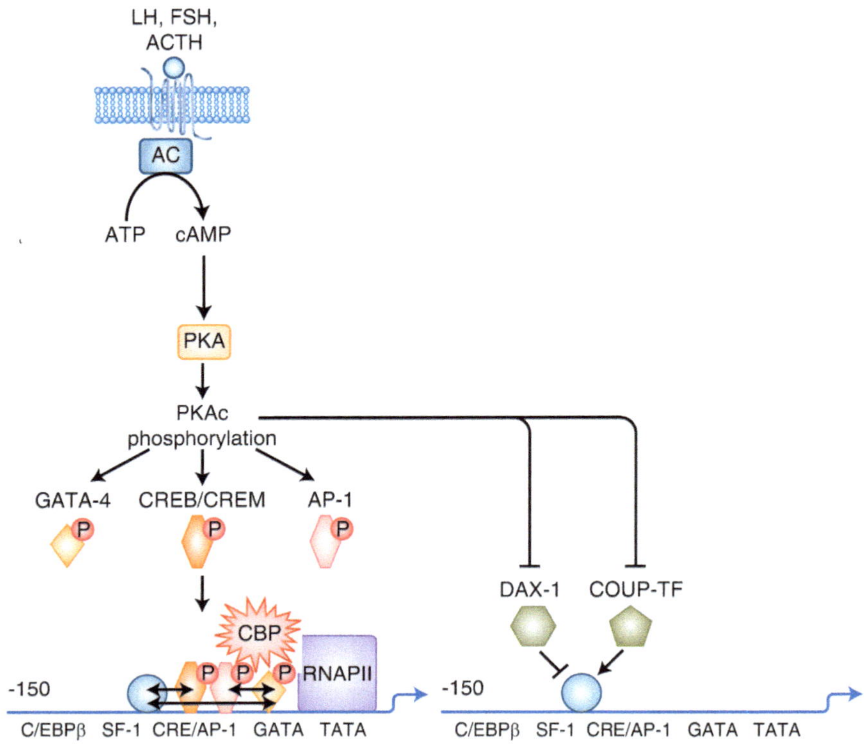

Fig. 2.5 Model for the cAMP-PKA-dependent activation of the mouse StAR gene. Tropic hormone (LH, FSH, ACTH) stimulation of their respective G-protein-coupled receptors (R) activates adenylyl cyclase (AC) and increases cAMP levels resulting in subsequent activation of protein kinase A (PKAc). PKAc-dependent phosphorylation of CREB/CREM, c-Fos/c-Jun (AP-1), and GATA-4 results in increased recruitment of these transcription factors to the StAR proximal promoter. The SF-1-GATA-4 binding appear to be a common mechanism for mouse and human StAR gene activation in multiple cell types while the bZIP homo- or heterodimer partners (CREB/CREM or AP-1 or CREB-cJun) binding to the CRE/AP-1 element is dependent upon the cell type and stimulus. Protein-protein interactions between GATA-4 and SF-1, AP-1 and/or CREB and possible CREB- SF-1 interactions help stabilize the transcription complex and enhance recruitment of the coactivator CBP and RNAPII to the StAR promoter and increase transcription. SF-1 may be constitutively bound to the promoter with increased association after cAMP-PKA activation. A switch from cAMP-dependent to cAMP-independent regulation of StAR transcription involves C/EBPb binding to its respective element and loss of CREB binding to the CRE/AP-1 element during luteolysis. NUR77 binds to a sequence overlapping the SF-1 element and modulates calcium-dependent regulation of StAR. Protein kinase A (PKAc)-dependent loss of DAX-1 and COUP-TF repressor functions: Both DAX-1 and COUP-TF function to interfere with SF-1 activation of the StAR promoter and actively repress StAR transcription (see text for details). Hormone stimulation and activation of PKAc results in decreased DAX-1 or COUP-TF expression in steroidogenic cells, thereby alleviating repression and promoting activation by allowing SF-1 and other factors to activate StAR gene expression *StAR* steroidogenic acute regulatory protein, *C/EBPb* CCAAT-enhancer-binding protein-beta, *CREB* cyclic-AMP responsive element-binding protein, *CREM* CREB modulator protein, *AP-1* activator protein 1, *cAMP* cyclic adenosine-3′,5′-monophosphate, *LH* luteinizing hormone, *FSH* follicule stimulating hormone, *ACTH* adrenocorticotropic hormone, *DAX-1* dosage sensitive sex reversal-adrenal hypoplasia congenital gene on the X chromosome gene 1, *COUP-TF* chicken ovalbumin upstream promoter transcription factor I and II, *CBP* CREB-binding protein, *RNAPII* ribonucleic acid polymerase II, bZIP basic leucine zipper, *AMP* adenosine-3′,5′-monophosphate

production switches from a cAMP-dependent (follicular phase) to cAMP-independent (luteal phase). StAR expression and steroid production are important during both phases and StAR gene regulation switches from a cAMP-dependent to a cAMP-independent mechanism. The switch is mediated by an AP-1 family member replacing CREB at the CRE/AP-1 element and recruitment of C/EBPβ to its functional element upstream of the CRE/AP-1 site [141]. ChIP analysis using mouse granulosa-luteal cells showed that C/EBPβ was recruited to the StAR promoter with no apparent increase in CREB/CREM association, indicating a tissue-specific difference in StAR regulation. These data are consistent with a switch in StAR regulation from a CREB/CREM-mediated (cAMP-dependent) to AP-1-C/EBPβ-mediated (cAMP-independent) mechanism in the ovary.

While much less is known about regulation of human StAR promoter activity, SF-1, GATA-4, and CREM remain important factors [78, 106–110]. The cAMP-dependent response in H295R human adrenocortical cells is CREM-mediated, and a CREM isoform has been shown to bind to the CRE/AP-1 element in the human StAR promoter [78, 110]. Overexpression of AP-1 family members JunB-Fos/FosB or JunB-ATF3 can transactivate StAR promoter-reporter gene expression in H295R cells, most likely via binding to the CRE/AP-1 element [105]. However, sequence differences in the CRE/AP-1 element between mouse and human StAR promoters may influence transcription factor binding, and this may broaden the transcription factor diversity for activating human StAR in response to different signaling pathways (Fig. 2.4b) [78].

Repression of StAR Transcription

In MA-10 mouse Leydig cells, a repressor complex composed of Sp3-mSin3A-HDAC proteins assembles on the mouse StAR promoter ~150 bp upstream of the CRE/AP-1 element [143]. Since the binding site for the repressor complex is not conserved, this may represent a species-(promoter)-specific repression of mouse StAR. In contrast, several repressor proteins have been shown to suppress StAR transcription by both direct and indirect mechanisms in several model systems and promoters. Repressor proteins can bind DNA and recruit corepressor complexes (direct), or block a trans-activator protein from binding DNA (indirect), thereby blocking assembly of a transcription initiation complex. The most extensively studied repressor for StAR is dosage sensitive sex reversal-adrenal hypoplasia congenital gene on the X chromosome gene 1 (Dax-1), a member of the nuclear receptor superfamily. DAX-1 is atypical in that it lacks a DNA-binding domain and contains a strong C-terminal repressor domain [144]. Expression patterns for DAX-1 and StAR in steroidogenic cell types are inversely related, providing correlative support for DAX-1 working as a repressor of StAR expression [145–153]. Despite the absence of a classical DNA-binding domain, DAX-1 has been reported to bind to a hairpin loop formed within promoters, including the StAR promoter [147, 154], which prevents SF-1 and/or NUR77 binding. In addition, direct SF-1-DAX-1 interactions that tether DAX-1 to the promoter via DNA-bound SF-1 results in recruitment of corepressor proteins and transcriptional repression of StAR [144, 154–156].

Chicken ovalbumin upstream promoter transcription factor I and II (COUP-TF), an orphan nuclear receptor, can function as either an activator or repressor protein [157]. As a repressor, COUP-TF can compete with SF-1 for binding to the AGGTCA element and recruit a corepressor complex, thus blocking SF-1-mediated transactivation [157–159]. COUP-TF can bind to SF-1 elements in the STAR promoter and overexpression of COUP-TF has been shown to block the hormone-stimulated increase in bovine StAR mRNA and human StAR promoter activity in bovine adrenal glomerulosa cells [160]. Thus, both DAX-1 and COUP-TF can function to interfere with SF-1 activation of the StAR promoter and represses StAR transcription. One possible mechanism for switching from repression to activation is hormone stimulation and activation of PKA resulting in decreased DAX-1 or COUP-TF expression, thereby allowing SF-1 and other factors to activate StAR gene expression (Fig. 2.5) [146, 153, 160, 161].

FOXL2, a member of the forkhead/hepatocyte nuclear factor 3 (FKH/HNF3) gene family, binds to the STAR proximal promoter (−42 bp) and represses StAR activity in the mouse ovary [162]. As with the other repressors, StAR and FOXL2 expression levels are inversely related; StAR mRNA levels are low in granulosa cells with high FOXL2. Differentiation of the follicle is associated with decreased FOXL2 and increased StAR, thus, hormone-dependent signaling appears to result in loss of a repressor and concomitant activation of transcription factors and coactivators to control StAR expression [141].

c-Fos and Yin yang 1 (YY1) have both activator and repressor functions on StAR promoter activity [78, 130, 163]. The rat StAR promoter has been well characterized for c-Fos, YY1, and DAX-1-mediated repression of StAR as a mechanism(s) contributing to suppression of steroid hormone biosynthesis and regression of the corpus luteum [130, 153, 163, 164]. Prostaglandin F2α (PGF2α) is a major signaling molecule for luteal regression in the rat ovary and PGF2α treatment was shown to increase c-Fos, YY1, and DAX-1 protein expression with subsequent repression of StAR promoter activity by both direct and indirect mechanisms.

Post-Transcriptional Regulation of StAR Expression

Two major StAR transcripts are detected for most mammalian species due to use of alternative polyadenylation sites in the 3′-UTR of the terminal exon [165] (Fig. 2.3). The differential processing of the 3′-end produces a short and long form of StAR mRNA with the long form retaining ~1800 nucleotides. The temporal pattern of expression of the two major transcripts after cAMP-PKA activation shows that both transcripts increase in parallel followed by a more rapid decrease in the long form relative to the short form [84, 85, 107, 166–168]. However, the steady-state mRNA levels, as determined by Northern blot analysis, indicates the short transcript is relatively minor compared to the long transcript, suggesting a preference for processing at the downstream polyadenylation site (84, 85, 107, 166–168]. The significance of the longer 3′UTR length is the presence of AU-rich destabilizing elements [169] that are commonly found in short-lived transcripts, consistent with the temporal patterns of expression observed with StAR long v short transcripts. The AURE-binding protein, TIS11b (zinc finger protein, Znf36L1, also known as BRF1), binds to the StAR AURE and facilitates mRNA degradation [170–172]. Classically, destabilizing element-binding proteins, such as TIS11b, function by recruiting de-capping and deadenylation enzymes to initiate mRNA degradation although the mechanism for StAR mRNA degradation remains to be determined. TIS11b expression in the adrenal and gonads is increased by cAMP-PKA [170], providing a mechanism for decreasing StAR steady-state mRNA levels as another control point to regulate StAR protein levels in response to hormonal stimulation [173].

In addition to mRNA stability, the 3′UTR of StAR mRNA may serve to localize the transcript to the mitochondria by binding to the A kinase anchor protein, 121 (AKAP 121; 174,175). AKAP 121 is a scaffold protein associated with mitochondria that binds PKA regulatory subunits thereby localizing PKA to this organelle [176]. PKA was shown to be part of a protein complex at the mitochondria outer membrane that also includes outer mitochondrial membrane translocator protein (TSPO), TSPO-associated protein (PAP7, an AKAP family member), and StAR[177]. The cAMP-dependent formation of this complex facilitates cholesterol transfer into the mitochondria and the presence of StAR is an absolute requirement for the function of this complex [177]. In the absence of AKAP121, StAR protein expression is diminished in MA-10 mouse Leydig cells, suggesting StAR mRNA localization to the outer mitochondrial membrane is important for efficient translation [97, 178]. Given that StAR mRNA may be recruited to the mitochondria by AKAP-121 and that AKAP-PKA associate at the mitochondria, this would allow for localized translation and posttranslational modification of StAR.

Summary

The journey from detecting four new protein spots by 2-D-PAGE after treatment of rat adrenal cells or MA-10 mouse Leydig tumor cells with cAMP analogs to activate PKA signaling to cloning the cDNA encoding the protein spans over a decade from the early 1980s to 1994. However, as with all research, the work over this time was building on seminal studies from the 1960s–1980s that provided the evidence for an acute regulator of steroidogenesis which should fit the following criteria: trophic hormone stimulation of steroidogenic cells should induce the rapid synthesis of a labile protein(s), and that this newly synthesized protein(s) should function at the site of mitochondria and facilitate the translocation of cholesterol across the mitochondrial membranes for delivery to the cytochrome P450 side chain cleavage enzyme which catalyzes the conversion of cholesterol to pregnenolone for the first enzymatic step in steroidogenesis. StAR fits this description and an impressive amount of work in the mid to late 1990s demonstrated that mutations in the StAR gene are the genetic basis for lipoid CAH, StAR knockout mice are unable to produce steroid hormones, and that StAR binds cholesterol and functions at the outer membrane of mitochondria for cholesterol transport into the organelle. Thus, relatively quickly, key data confirmed StAR's role as the acute regulator of steroidogenesis. Given the importance of the criteria for new protein synthesis for the acute regulation of steroidogenesis, understanding the molecular mechanisms that control the increase in StAR expression following trophic hormone stimulation would be important for understanding the mechanisms that control steroidogenesis. The cAMP responsive region of mouse and human StAR promoter is a non-consensus CRE/AP-1 element with overlapping specificity for members of the CREB/CREM and AP-1 protein families. Thus, the cAMP-PKA-dependent control of StAR was found to be rather complex and shown to occur in a species- and tissue-specific manner, driven mainly by the potential species- and tissue-specific heterodimer formations between bZIP CREB and AP-1 family members. Binding of other transcription factors, e.g., SF-1 and GATA-4, to their respective elements that are highly conserved between species may influence binding of the bZIP heterodimer partners at the non-consensus CRE/AP-1. Loss of transcriptional repression via loss of transcriptional repressor proteins DAX-1 and/or COUP-TF is another mechanism that works in concert with the direct activation mechanisms to promote the cAMP-PKA-dependent increase in StAR expression. Lastly, StAR mRNA stability is controlled by the AURE-binding protein, TIS11b, which binds to the StAR 3′ UTR AURE and facilitates mRNA degradation in a cAMP-PKA-dependent manner. Together these studies highlight the importance for fine control of StAR expression and that altering the expression or function of any of these transcription factors could influence StAR expression and lead to cell-specific alterations in steroid hormone biosynthesis, which would have implications for development and reproduction.

References

1. Macchi IA, Hechter O. Studies of ACTH action upon perfused bovine adrenals; duration of ACTH action. Endocrinology. 1954;55:434–8.
2. Macchi IA, Hechter O. Studies of ACTH action upon perfused bovine adrenals; minimal ACTH concentration requisite for maximal glandular response. Endocrinology. 1954;55:426–33.
3. Macchi IA, Hechter O. Studies of ACTH action upon perfused bovine adrenals; corticosteroid biosynthesis in isolated glands maximally stimulated with ACTH. Endocrinology. 1954;55:387–402.
4. Stone D, Hechter O. Studies on ACTH action in perfused bovine adrenals: the site of action of ACTH in corticosteroidogenesis. Arch Biochem Biophys. 1954;51:457–69.
5. Ferguson JJ, Jr. Puromycin and adrenal responsiveness to adrenocorticotropic hormone. Biochim Biophys Acta. 1962;57:616–7.
6. Ferguson JJ, Jr. Protein synthesis and adrenocorticotropin responsiveness. J Biol Chem. 1963;238:2754–9.
7. Garren LD, Davis WW, Crocco RM, Ney RL. Puromycin analogs: action of adrenocorticotropic hormone and the role of glycogen. Science. 1966;152:1386–8.
8. Garren LD, Gill GN, Masui H, Walton GM. On the mechanism of action of ACTH. Recent Prog Horm Res. 1971;27:433–78.
9. Garren LD, Ney RL, Davis WW. Studies on the role of protein synthesis in the regulation of corticosterone production by adrenocorticotropic hormone in vivo. Proc Natl Acad Sci U S A. 1965;53:1443–50.

10. Davis WW, Garren LD. Evidence for the stimulation by adenocorticotropic hormone of the conversion of cholesterol esters to cholesterol in the adrenal, in vivo. Biochem Biophys Res Commun. 1966;24:805–10.
11. Davis WW, Garren LD. On the mechanism of action of adrenocorticotropic hormone. The inhibitory site of cycloheximide in the pathway of steroid biosynthesis. J Biol Chem. 1968;243:5153–7.
12. Hall PF, Eik-Nes KB. The action of gonadotropic hormones upon rabbit testis in vitro. Biochim Biophys Acta. 1962;63:411–22.
13. Hall PF, Koritz S. Influence of interstitial cell-stimulating hormone on the conversion of cholesterol to progesterone by bovine corpus luteum. Biochemistry. 1965;4:1037–43.
14. Hall PF, Koritz S. Action of ACTH upon steroidogenesis in the chicken-adrenal gland. Endocrinology. 1966;79:652–4.
15. Koritz SB, Hall PF. Further studies on the locus of action of interstitial cell-stimulating hormone on the biosynthesis of progesterone by bovine corpus luteum. Biochemistry. 1965;4:2740–7.
16. Brownie AC, Simpson ER, Jefcoate CR, Boyd GS, Orme-Johnson WH, Beinert H. Effect of ACTH on cholesterol side-chain cleavage in rat adrenal mitochondria. Biochem Biophys Res Commun. 1972;46:483–90.
17. Paul DP, Gallant S, Orme-Johnson NR, Orme-Johnson WH, Brownie AC. Temperature dependence of cholesterol binding to cytochrome P-450scc of the rat adrenal. Effect of adrenocorticotropic hormone and cycloheximide. J Biol Chem. 1976;251:7120–6.
18. Crivello JF, Jefcoate CR. Mechanisms of corticotropin action in rat adrenal cells. I. The effects of inhibitors of protein synthesis and of microfilament formation on corticosterone synthesis. Biochim Biophys Acta. 1978;542:315–29.
19. Simpson ER, Trzeciak WH, McCarthy JL, Jefcoate CR, Boyd GS. Factors affecting cholesterol esterase and cholesterol side-chain-cleavage activities in rat adrenal. Biochem J. 1972;129:10P–11P.
20. Privalle CT, Crivello JF, Jefcoate CR. Regulation of intramitochondrial cholesterol transfer to side-chain cleavage cytochrome P-450 in rat adrenal gland. Proc Natl Acad Sci U S A. 1983;80:702–6.
21. Jefcoate CR, DiBartolomeis MJ, Williams CA, McNamara BC. ACTH regulation of cholesterol movement in isolated adrenal cells. J Steroid Biochem. 1987;27:721–9.
22. Farese RV. Inhibition of the steroidogenic effect of acth and incorporation of amino acid into rat adrenal protein in vitro by chloramphenicol. Biochim Biophys Acta. 1964;87:699–701.
23. Farese RV. Effects of actinomycin D on ACTH-induced corticosteroidogenesis. Endocrinology. 1966;78:929–36.
24. Farese RV. Adrenocorticotrophin-induced changes in the steroidogenic activity of adrenal cell-free preparations. Biochemistry. 1967;6:2052–65.
25. Farese RV, Prudente WJ. On the requirement for protein synthesis during corticotropin-induced stimulation of cholesterol side chain cleavage in rat adrenal mitochondrial and solubilized desmolase preparations. Biochim Biophys Acta. 1977;496:567–70.
26. Farese RV, Prudente WJ. On the role of intra-adrenal unesterified cholesterol in the steroidogenic effect of corticotropin. Biochim Biophys Acta. 1978;544:77–84.
27. Arthur JR, Boyd GS. The effect of inhibitors of protein synthesis on cholesterol side-chain cleavage in the mitochondria of luteinized rat ovaries. Eur J Biochem. 1974;49:117–27.
28. Robinson J, Stevenson PM, Boyd GS, Armstrong DT. Acute in vivo effects on HCG and LH on ovarian mitochondrial cholesterol utilization. Mol Cell Endocrinol. 1975;2:149–55.
29. Mason JI, Arthur JR, Boyd GS. Control of sterol metabolism in rat adrenal mitochondria. Biochem J. 1978;174:1045–51.
30. Simpson ER, McCarthy JL, Peterson JA. Evidence that the cycloheximide-sensitive site of adrenocorticotropic hormone action is in the mitochondrion. Changes in pregnenolone formation, cholesterol content, and the electron paramagnetic resonance spectra of cytochrome P-450. J Biolo Chem. 1978;253:3135–9.
31. Mason JI, Arthur JR, Boyd GS. Regulation of cholesterol metabolism in rat adrenal mitochondria. Mol Cell Endocrinol. 1978;10:209–23.

32. Haynes RC, Jr., Koritz SB, Peron FG. Influence of adenosine 3′,5′-monophosphate on corticoid production by rat adrenal glands. J Biol Chem. 1959;234:1421–3.
33. Karaboyas GC, Koritz S. The transformation of delta-5-pregnenolone and progesterone to cortisol by rat adrenal slices and the effect of acth and adenosine 3′-,5′-monophosphate upon it. Biochim Biophys Acta. 1965;100:600–2.
34. Cooke BA, Janszen FH, Clotscher WF, van der Molen HJ. Effect of protein-synthesis inhibitors on testosterone production in rat testis interstitial tissue and Leydig-cell preparations. Biochem J. 1975;150:413–8.
35. Cooke BA, Lindh LM, van der Molen HJ. The mechanism of action of lutropin on regulator protein(s) involved in Leydig-cell steroidogenesis. Biochem J. 1979;184:33–8.
36. Janszen FH, Cooke BA, van der Molen HJ. Specific protein synthesis in isolated rat testis leydig cells. Influence of luteinizing hormone and cycloheximide. Biochem J. 1977;162:341–6.
37. Janszen FH, Cooke BA, van Driel MJ, van der Molen HJ. The effect of lutropin on specific protein synthesis in tumour Leydig cells and in Leydig cells from immature rats. Biochem J. 1978;172:147–53.
38. Alberta JA, Epstein LF, Pon LA, Orme-Johnson NR. Mitochondrial localization of a phosphoprotein that rapidly accumulates in adrenal cortex cells exposed to adrenocorticotropic hormone or to cAMP. J Biol Chem. 1989;264:2368–72.
39. Epstein LF, Orme-Johnson NR. Regulation of steroid hormone biosynthesis. Identification of precursors of a phosphoprotein targeted to the mitochondrion in stimulated rat adrenal cortex cells. J Biol Chem. 1991;266:19739–45.
40. Epstein LF, Orme-Johnson NR. Acute action of luteinizing hormone on mouse Leydig cells: accumulation of mitochondrial phosphoproteins and stimulation of testosterone synthesis. Mol Cell Endocrinol. 1991;81:113–126.
41. Krueger RJ, Orme-Johnson NR. Acute adrenocorticotropic hormone stimulation of adrenal corticosteroidogenesis. Discovery of a rapidly induced protein. J Biol Chem. 1983;258:10159–67.
42. O'Farrell PH, O'Farrell PZ. Two-dimensional polyacrylamide gel electrophoretic fractionation. Methods Cell Biol. 1977;16:407–20.
43. O'Farrell PZ, Goodman HM, O'Farrell PH. High resolution two-dimensional electrophoresis of basic as well as acidic proteins. Cell. 1977;12:1133–41.
44. Pon LA, Hartigan JA, Orme-Johnson NR. Acute ACTH regulation of adrenal corticosteroid biosynthesis. Rapid accumulation of a phosphoprotein. J Biol Chem. 1986;261:13309–16.
45. Pon LA, Orme-Johnson NR. Acute stimulation of steroidogenesis in corpus luteum and adrenal cortex by peptide hormones. Rapid induction of a similar protein in both tissues. J Biol Chem. 1986;261:6594–9.
46. Ascoli M. Characterization of several clonal lines of cultured Leydig tumor cells: gonadotropin receptors and steroidogenic responses. Endocrinology. 1981;108:88–95.
47. Stocco DM. Further evidence that the mitochondrial proteins induced by hormone stimulation in MA-10 mouse Leydig tumor cells are involved in the acute regulation of steroidogenesis. J Steroid Biochem Mol Biol. 1992;43:319–33.
48. Stocco DM, Ascoli M. The use of genetic manipulation of MA-10 Leydig tumor cells to demonstrate the role of mitochondrial proteins in the acute regulation of steroidogenesis. Endocrinology. 1993;132:959–67.
49. Stocco DM, Chen W. Presence of identical mitochondrial proteins in unstimulated constitutive steroid-producing R2 C rat Leydig tumor and stimulated nonconstitutive steroid-producing MA-10 mouse Leydig tumor cells. Endocrinology. 1991;128:1918–26.
50. Stocco DM, Kilgore MW. Induction of mitochondrial proteins in MA-10 Leydig tumour cells with human choriogonadotropin. Biochem J. 1988;249:95–103.
51. Stocco DM, King S, Clark BJ. Differential effects of dimethylsulfoxide on steroidogenesis in mouse MA-10 and rat R2 C Leydig tumor cells. Endocrinology. 1995;136:2993–9.
52. Stocco DM, Sodeman TC. The 30-kDa mitochondrial proteins induced by hormone stimulation in MA-10 mouse Leydig tumor cells are processed from larger precursors. J Biol Chem. 1991;266:19731–8.

53. Clark BJ, Wells J, King SR, Stocco DM. The purification, cloning, and expression of a novel luteinizing hormone-induced mitochondrial protein in MA-10 mouse Leydig tumor cells. Characterization of the steroidogenic acute regulatory protein (StAR). J Biol Chem. 1994;269:28314–22.
54. Lin D, Gitelman SE, Saenger P, Miller WL. Normal genes for the cholesterol side chain cleavage enzyme, P450scc, in congenital lipoid adrenal hyperplasia. J Clin Invest. 1991;88:1955–62.
55. Lin D, Sugawara T, Strauss JF, 3rd, Clark BJ, Stocco DM, Saenger P, Rogol A, Miller WL. Role of steroidogenic acute regulatory protein in adrenal and gonadal steroidogenesis. Science. 1995;267:1828–31.
56. Caron KM, Soo SC, Wetsel WC, Stocco DM, Clark BJ, Parker KL. Targeted disruption of the mouse gene encoding steroidogenic acute regulatory protein provides insights into congenital lipoid adrenal hyperplasia. Proc Natl Acad Sci U S Am. 1997;94:11540–5.
57. Gucev ZS, Tee MK, Chitayat D, Wherrett DK, Miller WL. Distinguishing deficiencies in the steroidogenic acute regulatory protein and the cholesterol side chain cleavage enzyme causing neonatal adrenal failure. J Pediatr. 2013;162:819–22.
58. Sahakitrungruang T, Tee MK, Blackett PR, Miller WL. Partial defect in the cholesterol side-chain cleavage enzyme P450scc (CYP11A1) resembling nonclassic congenital lipoid adrenal hyperplasia. J Clin Endocrinol Metab. 2011;96:792–8.
59. Tee MK, Abramsohn M, Loewenthal N, Harris M, Siwach S, Kaplinsky A, Markus B, Birk O, Sheffield VC, Pavari R, Hershkovitz E, Miller WL. Varied clinical presentations of seven patients with mutations in CYP11A1 encoding the cholesterol side-chain cleavage enzyme, P450scc. J Clin Endocrinol Metab. 2013;98:713–20.
60. Stocco DM. StAR protein and the regulation of steroid hormone biosynthesis. Annu Rev Physiol. 2001;63:193–213.
61. King SR, Bhangoo A, Stocco DM. Functional and physiological consequences of StAR deficiency: role in lipoid congenital adrenal hyperplasia. Endocr Dev. 2011;20:47–53.
62. Arakane F, Sugawara T, Nishino H, Liu Z, Holt JA, Pain D, Stocco DM, Miller WL, Strauss JF, 3rd. Steroidogenic acute regulatory protein (StAR) retains activity in the absence of its mitochondrial import sequence: implications for the mechanism of StAR action. Proc Natl Acad Sci U S A. 93:13731–6.
63. Caron KM, Soo SC, Wetsel WC, Stocco DM, Clark BJ, Parker KL. Targeted disruption of the mouse gene encoding steroidogenic acute regulatory protein provides insights into congenital lipoid adrenal hyperplasia. Proc Natl Acad Sci U S A. 1997;94:11540–5.
64. Sasaki G, Ishii T, Jeyasuria P, Jo Y, Bahat A, Orly J, Hasegawa T, Parker KL. Complex role of the mitochondrial targeting signal in the function of steroidogenic acute regulatory protein revealed by bacterial artificial chromosome transgenesis in vivo. Mol Endocrinol. 2008;22:951–64.
65. Aoyagi S, Archer TK. Dynamics of coactivator recruitment and chromatin modifications during nuclear receptor mediated transcription. Mol Cell Endocrinol. 2008;280:1–5.
66. Lonard DM, O'Malley B W. Nuclear receptor coregulators: judges, juries, and executioners of cellular regulation. Mol Cell. 2007;27:691–700.
67. Rosenfeld MG, Lunyak VV, Glass CK. Sensors and signals: a coactivator/corepressor/epigenetic code for integrating signal-dependent programs of transcriptional response. Genes Dev. 2006;20:1405–28.
68. Savkur RS, Burris TP. The coactivator LXXLL nuclear receptor recognition motif. J Pep Res. 2004;63:207–12.
69. Xu L, Glass CK, Rosenfeld MG. Coactivator and corepressor complexes in nuclear receptor function. Curr Opin Genet, Dev. 1999;9:140–47.
70. Xu W, Kasper LH, Lerach S, Jeevan T, Brindle PK. Individual CREB-target genes dictate usage of distinct cAMP-responsive coactivation mechanisms. Embo J. 2007;26:2890–903.
71. Plevin MJ, Mills MM, Ikura M. The LxxLL motif: a multifunctional binding sequence in transcriptional regulation. Trends Biochem Sci. 2005;30:66–69.
72. Wysocka J, Allis CD, Coonrod S. Histone arginine methylation and its dynamic regulation. Front Biosci. 2006;11:344–55.

73. Kao HY, Downes M, Ordentlich P, Evans RM. Isolation of a novel histone deacetylase reveals that class I and class II deacetylases promote SMRT-mediated repression. Genes Dev. 2000;14:55–66.
74. Downes M, Ordentlich P, Kao HY, Alvarez JG, Evans RM. Identification of a nuclear domain with deacetylase activity. Proc Natl Acad Sci U S A. 2000;97:10330–5.
75. Alpy F, Rousseau A, Schwab Y, Legueux F, Stoll I, Wendling C, Spiegelhalter C, Kessler P, Mathelin C, Rio MC, Levine TP, Tomasetto C. STARD3/STARD3NL and VAP make a novel molecular tether between late endosomes and the ER. J Cell Sci. 2013;126:5500–12.
76. Hattangady NG, Olala LO, Bollag WB, Rainey WE. Acute and chronic regulation of aldosterone production. Mol Cell Endocrinol. 2012;350:151–62.
77. Li J, Feltzer RE, Dawson KL, Hudson EA, Clark BJ. Janus kinase 2 and calcium are required for angiotensin II-dependent activation of steroidogenic acute regulatory protein transcription in H295R human adrenocortical cells. J Biol Chem. 2003;278:52355–62.
78. Meier RK, Clark BJ. Angiotensin II-dependent transcriptional activation of human steroidogenic acute regulatory protein gene by a 25-kDa cAMP-responsive element modulator protein isoform and Yin Yang 1. Endocrinology. 2012;153:1256–68.
79. Nogueira EF, Bollag WB, Rainey WE. Angiotensin II regulation of adrenocortical gene transcription. Mol Cell Endocrinol. 2009;302:230–6.
80. Nogueira EF, Rainey WE. Regulation of aldosterone synthase by activator transcription factor/cAMP response element-binding protein family members. Endocrinology. 2010;151:1060–70.
81. Arakane F, King SR, Du Y, Kallen CB, Walsh LP, Watari H, Stocco DM, Strauss JF, 3rd. Phosphorylation of steroidogenic acute regulatory protein (StAR) modulates its steroidogenic activity. J Biol Chem. 1997;272:32656–62.
82. Baker BY, Epand RF, Epand RM, Miller WL. Cholesterol binding does not predict activity of the steroidogenic acute regulatory protein, StAR. J Biol Chem. 2007;282:10223–32.
83. Jo Y, King SR, Khan SA, Stocco DM. Involvement of protein kinase C and cyclic adenosine 3′,5′-monophosphate-dependent kinase in steroidogenic acute regulatory protein expression and steroid biosynthesis in Leydig cells. Biol Reprod. 2005;73:244–55.
84. Clark BJ, Soo SC, Caron KM, Ikeda Y, Parker KL, Stocco DM. Hormonal and developmental regulation of the steroidogenic acute regulatory protein. Mol Endocrinol. 1995;9:1346–55.
85. Sugawara T, Lin D, Holt JA, Martin KO, Javitt NB, Miller WL, Strauss JF, 3rd. Structure of the human steroidogenic acute regulatory protein (StAR) gene: StAR stimulates mitochondrial cholesterol 27-hydroxylase activity. Biochemistry. 1995;34:12506–12.
86. Honda S, Morohashi K, Nomura M, Takeya H, Kitajima M, Omura T. Ad4BP regulating steroidogenic P-450 gene is a member of steroid hormone receptor superfamily. J Biol Chem. 1993;268:7494–502.
87. Lala DS, Rice DA, Parker KL. Steroidogenic factor I, a key regulator of steroidogenic enzyme expression, is the mouse homolog of fushi tarazu-factor I. Mol Endocrinol. 1992;6:1249–58.
88. Luo X, Ikeda Y, Parker KL. A cell-specific nuclear receptor is essential for adrenal and gonadal development and sexual differentiation. Cell. 1994;77:481–90.
89. Morohashi K, Zanger UM, Honda S, Hara M, Waterman MR, Omura T. Activation of CYP11A and CYP11B gene promoters by the steroidogenic cell-specific transcription factor, Ad4BP. Mol Endocrinol. 1993;7:1196–204.
90. Hoivik EA, Bjanesoy TE, Bakke M. Epigenetic regulation of the gene encoding steroidogenic factor-1. Mol Cell Endocrinol. 2013;371:133–9.
91. Hoivik EA, Witsoe SL, Bergheim IR, Xu Y, Jakobsson I, Tengholm A, Doskeland SO, Bakke M. DNA methylation of alternative promoters directs tissue specific expression of epac2 isoforms. PLoS One. 2013;8:e67925.
92. Parker KL, Rice DA, Lala DS, Ikeda Y, Luo X, Wong M, Bakke M, Zhao L, Frigeri C, Hanley NA, Stallings N, Schimmer BP. Steroidogenic factor. 1: an essential mediator of endocrine development. Recent Prog Horm Res. 2002;57:19–36.
93. Xu B, Yang W-H, Gerin I, Hu C-D, Hammer GD, Koenig RJ. Dax-1 and steroid receptor RNA activator (SRA) function as transcriptional coactivators for steroidogenic factor 1 in steroidogenesis. Mol Cell Biol. 2009;29:1719–34.

94. Schimmer BP, White PC. Minireview: steroidogenic factor. 1: its roles in differentiation, development, and disease. Mol Endocrinol. 2010;24:1322–37.
95. Giguere V. Orphan nuclear receptors: from gene to function. Endocr Rev. 1999;20:689–725.
96. Davis IJ, Lau LF. Endocrine and neurogenic regulation of the orphan nuclear receptors Nur77 and Nurr-1 in the adrenal glands. Mol Cell Biol. 1994;14:3469–83.
97. Li W, Amri H, Huang H, Wu C, Papadopoulos V. Gene and protein profiling of the response of MA-10 Leydig tumor cells to human chorionic gonadotropin. J Androl. 2004;25:900–13.
98. Martin LJ, Tremblay JJ. The human 3beta-hydroxysteroid dehydrogenase/Delta5-Delta4 isomerase type 2 promoter is a novel target for the immediate early orphan nuclear receptor Nur77 in steroidogenic cells. Endocrinology. 2005;146:861–9.
99. Bassett MH, Suzuki T, Sasano H, White PC, Rainey WE. The orphan nuclear receptors NURR1 and NGFIB regulate adrenal aldosterone production. Mol Endocrinol. 2004;18:279–90.
100. Havelock JC, Smith AL, Seely JB, Dooley CA, Rodgers RJ, Rainey WE, Carr BR. The NGFI-B family of transcription factors regulates expression of 3beta-hydroxysteroid dehydrogenase type 2 in the human ovary. Mol Hum Reprod. 2005;11:79–85.
101. Martin LJ, Taniguchi H, Robert NM, Simard J, Tremblay JJ, Viger RS. GATA factors and the nuclear receptors, steroidogenic factor 1/liver receptor homolog 1, are key mutual partners in the regulation of the human 3beta-hydroxysteroid dehydrogenase type 2 promoter. Mol Endocrinol. 2005;19:2358–70.
102. Martin LJ, Tremblay JJ. The nuclear receptors NUR77 and SF1 play additive roles with c-JUN through distinct elements on the mouse Star promoter. J Mol Endocrinol. 2009;42:119–29.
103. Abdou HS, Villeneuve G, Tremblay JJ. The calcium signaling pathway regulates leydig cell steroidogenesis through a transcriptional cascade involving the nuclear receptor NR4A1 and the steroidogenic acute regulatory protein. Endocrinology. 2013;154:511–20.
104. Martin LJ, Boucher N, Brousseau C, Tremblay JJ. The orphan nuclear receptor NUR77 regulates hormone-induced StAR transcription in Leydig cells through cooperation with Ca2+/calmodulin-dependent protein kinase I. Mol Endocrinol. 2008;22:2021–37.
105. Nogueira EF, Xing Y, Morris CA, Rainey WE. Role of angiotensin II-induced rapid response genes in the regulation of enzymes needed for aldosterone synthesis. J Mol Endocrinol. 2009;42:319–30.
106. Sugawara T, Holt JA, Kiriakidou M, Strauss JF, 3rd. Steroidogenic factor 1-dependent promoter activity of the human steroidogenic acute regulatory protein (StAR) gene. Biochemistry. 1996;35:9052–9.
107. Clark BJ, Combs R. Angiotensin II and cyclic adenosine 3′,5′-monophosphate induce human steroidogenic acute regulatory protein transcription through a common steroidogenic factor-1 element. Endocrinology. 1999;140:4390–8.
108. Sugawara T, Kiriakidou M, McAllister JM, Kallen CB, Strauss JF, 3rd. Multiple steroidogenic factor 1 binding elements in the human steroidogenic acute regulatory protein gene 5′-flanking region are required for maximal promoter activity and cyclic AMP responsiveness. Biochemistry. 1997;36:7249–55.
109. Sugawara T, Saito M, Fujimoto S. Sp1 and SF-1 interact and cooperate in the regulation of human steroidogenic acute regulatory protein gene expression. Endocrinology. 2000;141:2895–903.
110. Sugawara T, Sakuragi N, Minakami H. CREM confers cAMP responsiveness in human steroidogenic acute regulatory protein expression in NCI-H295R cells rather than SF-1/Ad4BP. J Endocrinol. 2006;191:327–37.
111. Arceci RJ, King AA, Simon MC, Orkin SH, Wilson DB. Mouse GATA-4: a retinoic acid-inducible GATA-binding transcription factor expressed in endodermally derived tissues and heart. Mol Cell Biol. 1993;13:2235–46.
112. Molkentin JD. The zinc finger-containing transcription factors GATA-4, -5, and -6. Ubiquitously expressed regulators of tissue-specific gene expression. J Biol Chem. 2000;275:38949–52.
113. Zaytouni T, Efimenko EE, Tevosian S. GATA transcription factors in the developing reproductive system. Adv Genet. 2011;76:93–134.

114. Manna PR, Dyson MT, Stocco DM. Regulation of the steroidogenic acute regulatory protein gene expression: present and future perspectives. Mol Hum Reprod. 2009;15:321–33.
115. Manna PR, Dyson MT, Stocco DM. Role of basic leucine zipper proteins in transcriptional regulation of the steroidogenic acute regulatory protein gene. Mol Cell Endocrinol. 2009;302:1–11.
116. Murphy NC, Biankin AV, Millar EK, McNeil CM, O'Toole SA, Segara D, Crea P, Olayioye MA, Lee CS, Fox SB, Morey AL, Christie M, Musgrove EA, Daly RJ, Lindeman GJ, Henshall SM, Visvader JE, Sutherland RL. Loss of STARD10 expression identifies a group of poor prognosis breast cancers independent of HER2/Neu and triple negative status. Int J Cancer. 2009;126:1445–53.
117. Altarejos JY, Montminy M. CREB and the CRTC co-activators: sensors for hormonal and metabolic signals. Nat Rev Mol Cell Biol. 2011;12:141–51.
118. Della Fazia MA, Servillo G, Sassone-Corsi P. Cyclic AMP signalling and cellular proliferation: regulation of CREB and CREM. FEBS Lett. 1997;410:22–4.
119. Hummler E, Cole TJ, Blendy JA, Ganss R, Aguzzi A, Schmid W, Beermann F, Schutz G. Targeted mutation of the CREB gene: compensation within the CREB/ATF family of transcription factors. Proc Natl Acad Sci U S A. 1994;91:5647–51.
120. Meyer TE, Habener JF. Cyclic AMP response element binding protein CREB and modulator protein CREM are products of distinct genes. Nucleic Acids Res. 1992;20:6106.
121. Sassone-Corsi P. CREM: a master-switch regulating the balance between differentiation and apoptosis in male germ cells. Mol Reprod Dev. 2000;56:228–9.
122. Clem BF, Hudson EA, Clark BJ. Cyclic adenosine 3′,5′-monophosphate (cAMP) enhances cAMP-responsive element binding (CREB) protein phosphorylation and phospho-CREB interaction with the mouse steroidogenic acute regulatory protein gene promoter. Endocrinology. 2005;146:1348–56.
123. Manna PR, Dyson MT, Eubank DW, Clark BJ, Lalli E, Sassone-Corsi P, Zeleznik AJ, Stocco DM. Regulation of steroidogenesis and the steroidogenic acute regulatory protein by a member of the cAMP response-element binding protein family. Mol Endocrinol. 2002;16:184–99.
124. Angel P, Karin M. The role of Jun, Fos and the AP-1 complex in cell-proliferation and transformation. Biochim Biophys Acta 1991;1072:129–57.
125. Hai T, Curran T. Cross-family dimerization of transcription factors Fos/Jun and ATF/CREB alters DNA binding specificity. Proc Natl Acad Sciences U S A. 1991;88:3720–4.
126. O'Shea EK, Rutkowski R, Kim PS. Mechanism of specificity in the Fos-Jun oncoprotein heterodimer. Cell. 1992;68:699–708.
127. Manna PR, Eubank DW, Stocco DM. Assessment of the role of activator protein-1 on transcription of the mouse steroidogenic acute regulatory protein gene. Mol Endocrinol. 2004;18:558–73.
128. Manna PR, Stocco DM. Crosstalk of CREB and Fos/Jun on a single cis-element: transcriptional repression of the steroidogenic acute regulatory protein gene. J Mol Endocrinol. 2007;39:261–77.
129. Manna PR, Stocco DM. The role of JUN in the regulation of PRKCC-mediated STAR expression and steroidogenesis in mouse Leydig cells. J Mol Endocrinol. 2008;41:329–41.
130. Shea-Eaton W, Sandhoff TW, Lopez D, Hales DB, McLean MP. Transcriptional repression of the rat steroidogenic acute regulatory (StAR) protein gene by the AP-1 family member c-Fos. Mol Cell Endocrinol. 2002;188:161–70.
131. Rincon Garriz JM, Suarez C, Capponi AM. c-Fos mediates angiotensin ii-induced aldosterone production and protein synthesis in bovine adrenal glomerulosa cells. Endocrinology. 2009;150:1294–302.
132. Osada S, Yamamoto H, Nishihara T, Imagawa M. DNA binding specificity of the CCAAT/enhancer-binding protein transcription factor family. J Biol Chem. 1996;271:3891–6.
133. Grimm SL, Rosen JM. The role of C/EBPbeta in mammary gland development and breast cancer. J Mammary Gland Biol Neoplasia. 2003;8:191–204.
134. Lekstrom-Himes J, Xanthopoulos KG. Biological role of the CCAAT/enhancer-binding protein family of transcription factors. J Biol Chemi. 1998;273:28545–8.

135. Ramji DP, Foka P. CCAAT/enhancer-binding proteins: structure, function and regulation. Biochem J. 2002;365:561–75.
136. Nalbant D, Williams SC, Stocco DM, Khan S. Luteinizing hormone-dependent gene regulation in Leydig cells may be mediated by CCAAT/enhancer-binding protein-β. Endocrinology. 1998;139:272–9.
137. Silverman E, Eimerl S, Orly J. CCAAT enhancer-binding protein beta and GATA-4 binding regions within the promoter of the steroidogenic acute regulatory protein (StAR) gene are required for transcription in rat ovarian cells. J Biol Chem. 1999;274:17987–96.
138. Silverman E, Yivgi-Ohana N, Sher N, Bell M, Eimerl S, Orly J. Transcriptional activation of the steroidogenic acute regulatory protein (StAR) gene: GATA-4 and CCAAT/enhancer-binding protein beta confer synergistic responsiveness in hormone-treated rat granulosa and HEK293 cell models. Mol Cell Endocrinol. 2006;252:92–101.
139. Reinhart AJ, Williams SC, Clark BJ, Stocco DM. SF-1 (steroidogenic factor-1) and C/EBP beta (CCAAT/enhancer binding protein-beta) cooperate to regulate the murine StAR (steroidogenic acute regulatory) promoter. Mol Endocrinol. 1999;13:729–41.
140. Tremblay JJ, Hamel F, Viger RS. Protein kinase A-dependent cooperation between GATA and CCAAT/enhancer-binding protein transcription factors regulates steroidogenic acute regulatory protein promoter activity. Endocrinology. 2002;143:3935–45.
141. Yivgi-Ohana N, Sher N, Melamed-Book N, Eimerl S, Koler M, Manna PR, Stocco DM, Orly J. Transcription of steroidogenic acute regulatory protein in the rodent ovary and placenta: alternative modes of cyclic adenosine 3′, 5′-monophosphate dependent and independent regulation. Endocrinology. 2009;150:977–89.
142. Hiroi H, Christenson LK, Chang L, Sammel MD, Berger SL, Strauss JF, 3rd. Temporal and spatial changes in transcription factor binding and histone modifications at the steroidogenic acute regulatory protein (stAR) locus associated with stAR transcription. Mol Endocrinol. 2004;18:791–806.
143. Clem BF, Clark BJ. Association of the mSin3A-histone deacetylase 1/2 corepressor complex with the mouse steroidogenic acute regulatory protein gene. Mol Endocrinol. 2006;20:100–113.
144. Ito M, Yu R, Jameson JL. DAX-1 inhibits SF-1-mediated transactivation via a carboxy-terminal domain that is deleted in adrenal hypoplasia congenita. Mol Cell Biol. 1997;17:1476–83.
145. Ahn SW, Gang G-T, Kim YD, Ahn R-S, Harris RA, Lee C-H, Choi H-S. Insulin directly regulates steroidogenesis via induction of the orphan nuclear receptor DAX-1 in testicular Leydig cells. J Biol Chem. 2013;288:15937–46.
146. Jo Y, Stocco DM. Regulation of Steroidogenesis and steroidogenic acute regulatory protein in R2 C cells by DAX-1 (dosage-sensitive sex reversal, adrenal hypoplasia congenita, critical region on the X chromosome, gene-1). Endocrinology. 2004;145:5629–37.
147. Manna PR, Dyson MT, Jo Y, Stocco DM. Role of dosage-sensitive sex reversal, adrenal hypoplasia congenita, critical region on the X chromosome, gene 1 in protein kinase A- and protein kinase C-mediated regulation of the steroidogenic acute regulatory protein expression in mouse Leydig tumor cells: mechanism of action. Endocrinology. 2009;150:187–99.
148. Shimizu T, Sudo N, Yamashita H, Murayama C, Miyazaki H, Miyamoto A. Histone H3 acetylation of StAR and decrease in DAX-1 is involved in the luteinization of bovine granulosa cells during in vitro culture. Mol Cell Biochem. 2009;328:41–7.
149. Li W, Pandey AK, Yin X, Chen JJ, Stocco DM, Grammas P, Wang X. Effects of apigenin on steroidogenesis and steroidogenic acute regulatory gene expression in mouse Leydig cells. J Nutr Biochem. 2011;22:212–8.
150. Pandey AK, Li W, Yin X, Stocco DM, Grammas P, Wang X. Blocking L-type calcium channels reduced the threshold of cAMP-induced steroidogenic acute regulatory gene expression in MA-10 mouse Leydig cells. J Endocrinol. 2010;204:67–74.
151. Yu C-C, Li P-H. In vivo inhibition of steroidogenic acute regulatory protein expression by dexamethasone parallels induction of the negative transcription factor DAX-1. Endocrine. 2006;30:313–23.
152. Ragazzon B, Lefrançois-Martinez A-M, Val P, Sahut-Barnola I, Tournaire C, Chambon C, Gachancard-Bouya J-L, Begue R-J, Veyssière G, Martinez A. Adrenocorticotropin-dependent

changes in SF-1/DAX-1 ratio influence steroidogenic genes expression in a novel model of glucocorticoid-producing adrenocortical cell lines derived from targeted tumorigenesis. Endocrinology. 2006;147:1805–18.
153. Sandhoff TW, McLean MP. Repression of the rat steroidogenic acute regulatory (StAR) protein gene by PGF2alpha is modulated by the negative transcription factor DAX-1. Endocrine. 1999;10:83–91.
154. Zazopoulos E, Lalli E, Stocco DM, Sassone-Corsi P. DNA binding and transcriptional repression by DAX-1 blocks steroidogenesis. Nature. 1997;390:311–5.
155. Tajima K, Dantes A, Yao Z, Sorokina K, Kotsuji F, Seger R, Amsterdam A. Down-regulation of steroidogenic response to gonadotropins in human and rat preovulatory granulosa cells involves mitogen-activated protein kinase activation and modulation of DAX-1 and steroidogenic factor-1. J Clin Endocrinol Metab. 2003;88:2288–99.
156. Ehrlund A, Anthonisen EH, Gustafsson N, Venteclef N, Robertson Remen K, Damdimopoulos AE, Galeeva A, Pelto-Huikko M, Lalli E, Steffensen KR, Gustafsson J-Å, Treuter E. E3 ubiquitin ligase RNF31 cooperates with DAX-1 in transcriptional repression of steroidogenesis. Mol Cell Biol. 2009;29:2230–42.
157. Shibata H, Kurihara I, Kobayashi S, Yokota K, Suda N, Saito I, Saruta T. Regulation of differential COUP-TF-coregulator interactions in adrenal cortical steroidogenesis. J Steroid Biochem Mol Biol. 2003;85:449–56.
158. Shibata H, Ikeda Y, Mukai T, Morohashi K, Kurihara I, Ando T, Suzuki T, Kobayashi S, Murai M, Saito I, Saruta T. Expression profiles of COUP-TF, DAX-1, and SF-1 in the human adrenal gland and adrenocortical tumors: possible implications in steroidogenesis. Mol Genet Metab. 2001;74:206–16.
159. Bakke M, Lund J. Mutually exclusive interactions of two nuclear orphan receptors determine activity of a cyclic adenosine 3′,5′-monophosphate-responsive sequence in the bovine CYP17 gene. Mol Endocrinol. 1995;9:327–39.
160. Buholzer CF, Arrighi JF, Abraham S, Piguet V, Capponi AM, Casal AJ. Chicken ovalbumin upstream promoter-transcription factor (COUP-TF) is a negative regulator of steroidogenesis in bovine adrenal glomerulosa cells. Mol Endocrinol. 2004;19:65–75.
161. Osman H, Murigande C, Nadakal A, Capponi AM. Repression of DAX-1 and induction of SF-1 expression. Two mechanisms contributing to the activation of aldosterone biosynthesis in adrenal glomerulosa cells. J Biol Chem. 2002;277:41259–67.
162. Pisarska MD, Bae J, Klein C, Hsueh AJ. Forkhead L2 is expressed in the ovary and represses the promoter activity of the steroidogenic acute regulatory gene. Endocrinology. 2004;145:3424–33.
163. Nackley AC, Shea-Eaton W, Lopez D, McLean MP. Repression of the steroidogenic acute regulatory gene by the multifunctional transcription factor Yin Yang 1. Endocrinology. 2002;143:1085–96.
164. Liu Q, Merkler KA, Zhang X, McLean MP. Prostaglandin F2{alpha} suppresses rat steroidogenic acute regulatory protein expression via induction of Yin Yang 1 protein and recruitment of histone deacetylase 1 protein. Endocrinology. 2007;148:5209–19.
165. Ariyoshi N, Kim YC, Artemenko I, Bhattacharyya KK, Jefcoate CR. Characterization of the rat Star gene that encodes the predominant 3.5-kilobase pair mRNA. ACTH stimulation of adrenal steroids in vivo precedes elevation of Star mRNA and protein. J Biol Chem. 1998;273:7610–19.
166. Sandhoff TW, McLean MP. Hormonal regulation of steroidogenic acute regulatory (StAR) protein messenger ribonucleic acid expression in the rat ovary. Endocrine. 1996;4:259–67.
167. Cherradi N, Brandenburger Y, Rossier MF, Vallotton MB, Stocco DM, Capponi AM. Atrial natriuretic peptide inhibits calcium-induced steroidogenic acute regulatory protein gene transcription in adrenal glomerulosa cells. Mol Endocrinol. 1998;12:962–72.
168. Hartung S, Rust W, Balvers M, Ivell R. Molecular cloning and in vivo expression of the bovine steroidogenic acute regulatory protein. Biochem Biophys Res Commun. 1995;215:646–53.
169. Burrows JAJ, Willis LK, Perlmutter DH. Chemical chaperones mediate increased secretion of mutant α1-antitrypsin (α1-AT) Z: a potential pharmacological strategy for prevention of liver injury and emphysema in α1-AT deficiency. Proc Natl Acad Sci U S A. 2000;97:1796–801.

170. Duan H, Cherradi N, Feige JJ, Jefcoate C. cAMP-dependent posttranscriptional regulation of steroidogenic acute regulatory (STAR) protein by the zinc finger protein ZFP36L1/TIS11b. Mol Endocrinol. 2009;23:497–509.
171. Duan H, Jefcoate CR. The predominant cAMP-stimulated 3 × 5 kb StAR mRNA contains specific sequence elements in the extended 3′UTR that confer high basal instability. J Mol Endocrinol. 2007;38:159–79.
172. Zhao D, Duan H, Kim YC, Jefcoate CR. Rodent StAR mRNA is substantially regulated by control of mRNA stability through sites in the 3′-untranslated region and through coupling to ongoing transcription. J Steroid Biochem Mol Biol. 2005;96:155–73.
173. Jefcoate CR, Lee J, Cherradi N, Takemori H, Duan H. cAMP stimulation of StAR expression and cholesterol metabolism is modulated by co-expression of labile suppressors of transcription and mRNA turnover. Mol Cell Endocrinol. 2011;336:53–62.
174. Grozdanov PN, Stocco DM. Short RNA molecules with high binding affinity to the KH motif of A-kinase anchoring protein 1 (AKAP1): implications for the regulation of steroidogenesis. Mol Endocrinol. 2012;26:2104–17.
175. Ginsberg MD, Feliciello A, Jones JK, Avvedimento EV, Gottesman ME. PKA-dependent Binding of mRNA to the Mitochondrial AKAP121 Protein. J Mol Biol. 2003;327:885–97.
176. Feliciello A, Gottesman ME, Avvedimento EV. The biological functions of A-kinase anchor proteins. J Mol Biol. 2001;308:99–114.
177. Liu J, Rone MB, Papadopoulos V. Protein-protein interactions mediate mitochondrial cholesterol transport and steroid biosynthesis. J Biol Chem. 2006;281:38879–93.
178. Dyson MT, Jones JK, Kowalewski MP, Manna PR, Alonso M, Gottesman ME, Stocco DM. Mitochondrial A-kinase anchoring protein 121 binds type II protein kinase A and enhances steroidogenic acute regulatory protein-mediated steroidogenesis in MA-10 mouse leydig tumor cells. Biol Reprod. 2008;78:267–77.

Chapter 3
START Domain Protein Structure and Ligand Specificity

Danny Létourneau, Pierre Lavigne, Andrée Lefebvre and Jean-Guy LeHoux

Abstract The steroidogenic acute regulatory protein (StAR)-related lipid transfer (START) domain is an α-helix/β-grip fold module of approximately 210 amino acids that binds a broad variety of lipids and sterols. START domain proteins have various expression patterns and cellular localizations and are involved in lipid metabolism, lipid transfer, and cell signalling. The α-helix/β-grip tertiary structure delimits an internal cavity forming the binding site. However, the determinants that dictate ligand specificity and the mechanism of ligand entry and exit are ill-defined. Herein, we review and discuss the current knowledge on ligand specificity and reversible binding mechanism by START domains from a structural, dynamical, and thermodynamical viewpoint. More specifically, we highlight that the START domains involved in sterol (STARD1, STARD3, STARD4, STARD5, and STARD6) and lipid (STARD2 and STARD11) binding have conserved residues that play a structural role in the formation of a cavity that fits the shape of their specific ligands. We also expose a correlation between structural fluctuations reported to occur in secondary and tertiary structure elements of START domains and we suggest conformational changes to allow for entry and exit of ligands.

From the START: Historical Background

For many years, Prof. Jean-Guy LeHoux's research group has been studying the control of mineralo- and gluco-corticoids biosynthesis both in vivo and in vitro. Among other achievements, they have shown that in animals on a low sodium diet,

J.-G. LeHoux, Ph.D. (✉) · P. Lavigne, Ph.D. · D. Létourneau, Ph.D. · A. Lefebvre, Ph.D.
Département de Biochimie, Faculté de médecine et des sciences de la santé,
Université de Sherbrooke, 3001, 12e avenue Nord, Sherbrooke, QC, J1H 5N4, Canada
e-mail: jean-guy.lehoux@usherbrooke.ca

P. Lavigne
e-mail: pierre.lavigne@usherbrooke.ca

D. Létourneau
e-mail: Danny.Letourneau@usherbrooke.ca

A. Lefebvre
e-mail: andree.lefebvre@usherbrooke.ca

B. J. Clark, D. M. Stocco (eds.), *Cholesterol Transporters of the START Domain Protein Family in Health and Disease,* DOI 10.1007/978-1-4939-1112-7_3,

angiotensin II increases the transformation of cholesterol to pregnenolone in adrenals [1] and that adrenocorticotropic hormone (ACTH) also enhances this conversion [2].

Early studies had already suggested that the conversion of cholesterol to pregnenolone, by the cytochrome P450 side chain cleavage enzyme system (P450scc), was the rate-limiting step of the steroidogenesis pathway [3, 4]. It was subsequently established that newly synthesized and unstable (short-lived) proteins mediated the acute steroidogenic response upon ACTH stimulation [3, 5, 6], without affecting P450scc activity [7]. Hence, the transport of cholesterol into mitochondria, and not P450scc activity, was thus shown to be the rate-limiting step in the metabolism of cholesterol to pregnenolone. Subsequent work by the group of Orme-Johnson covering the period of 1983–1991 clearly established the existence of a family of phosphoproteins, ranging between 28–32 kDa in size, fulfilled the requirements of the postulated labile steroidogenic stimulatory protein in various steroidogenic tissues [8, 9]. In 1994, purification of a 30 kDa proteins from MA-10 cells by Clark et al. [10] allowed the cloning of its complementary deoxyribonucleic acid (cDNA). These authors found that this novel protein was required in the acute regulation of steroidogenesis and they proposed to name it steroidogenic acute regulatory (StAR, herein named STARD1) protein. Lin et al. [11] reported a mutated and non-functional STARD1 in three individuals with lipoid congenital adrenal hyperplasia, a disorder that is characterized by impaired gonadal and adrenal steroidogenesis. This salient work ignited our interest in the quest of a detailed understanding of the molecular and structural basis of the function and dysfunction of STARD1.

The cloning of STARD1 cDNA was subsequently reported for other species, including the hamster by our group [12]. Indirect immunofluorescence of adrenal paraffin sections revealed the presence of STARD1 in the adrenal *zonae glomerulosa, fasciculata,* and *reticularis*. The level of STARD1 messenger ribonucleic acid (mRNA) in both adrenal zones were increased within minutes after ACTH administration in animals [12] indicating that in vivo ACTH regulates STARD1 mRNA synthesis. Colloidal-gold electron microscopic studies on thin rat adrenal sections revealed that more than 98 % of positive STARD1 labeling was located over mitochondria [13]. Like in human, removal of the hamster STARD1 mitochondrial import sequence does not affect steroidogenesis [14], indicating that STARD1 can act on the outer mitochondrial membrane. STARD1 phosphorylation is essential for its activity. STARD1 transected COS-1 cells were incubated with [^{33}P]orthophosphate; following immunoprecipitation and immunoblotting, we found two ^{33}P-labeled protein bands migrating at the same level as STARD1 [14]. After digestion of these bands with hydrochloric acid (HCl) and separation by chromatography, STARD1 was found phosphorylated mainly on serine residues and to a much lower extent on threonine residues. Directed mutagenesis, immunoblotting with an anti-phospho-(S/T), and STARD1 activity experiments revealed that serine residues 194, 55, and 56 (in hamster) are important phosphorylation sites. In parallel and in order to get a deeper understanding of the specificity and mechanism of ligand (sterol) binding and dissociation, we initiated, 14 years ago, the detailed characterization of the structure, the dynamics, and thermodynamics of START domain proteins in absence and presence of ligands. With Prof. Pierre Lavigne, we joined our efforts to solve the structure and the mechanism of cholesterol binding of STARD1, STARD5, and STARD6. We reported the first 3D homology model of STARD1 [15] based on the crystal structure of metastatic lymph node 64 protein (MLN-64; STARD3; [16]) and revealed the presence of an internal cavity with the shape of one cholesterol molecule. Ten years after, this model was proven to be almost identical with the crystal structure recently solved by Thorsell et al. [17]. Using structural thermodynamical calculations, we proposed that in absence of ligand, the water-filled cavity of STARD1 (and other START domains) would promote the local unfolding of the C-terminal helix and hence the population of a partially folded state that would allow access of cholesterol to the internal cavity otherwise inaccessible. As detailed in the chapter, this model was subsequently supported by thermodynamical and functional studies from two different laboratories (ours and Walter Miller's) on STARD1 mutants designed to promote or reduce the opening of the STARD1 tertiary structure. Shortly afterwards and based on steered molecular dynamics simulation, an alternative model was also proposed by Murcia et al. [18]. According to this model, structural fluctuations in the Omega-1 loop are suggested to give access and allow exit from the internal cavity of STARD1 and STARD3. To date, no irrefutable data exist to conclude on the exact mechanism of ligand entry and exit from the cavity of START domains for the STARD1 and STARD4 subfamilies. With Dr. Danny Létourneau, we are presently completing the structure, dynamics, and thermodynamical analysis of STARD1, STARD5, and STARD 6 in solution using nuclear magnetic resonance spectroscopy in order to provide more conclusive experimental insights into the mechanism of ligand binding.

Introduction

The steroidogenic acute regulatory protein (StAR)-related lipid transfer (START) domain is an α-helix/β-grip fold module of approximately 210 amino acids [16, 19]; START domain proteins bind a wide variety of lipids and sterols [20]. The α-helix/β-grip tertiary structure delimits an internal cavity forming the binding site (Fig. 3.1a; [21]). Absent in yeast and archaea, START domain proteins are expressed in plants and animals and are conserved through evolution. In plants, START domain proteins are often found in homeodomains and thus play a role in gene expression [20]. In vertebrates, multi-domain proteins containing a START domain are frequent and are involved in protein localization, enzymatic activity, and cell signalling [19, 22]. In humans, 15 different proteins possess a START domain [23] and are divided in six subfamilies [24]. These proteins have various expression patterns and cellular localizations and are involved in lipid metabolism, lipid transfer, and cell signalling. The cellular functions and ligands of the mammalian START containing proteins

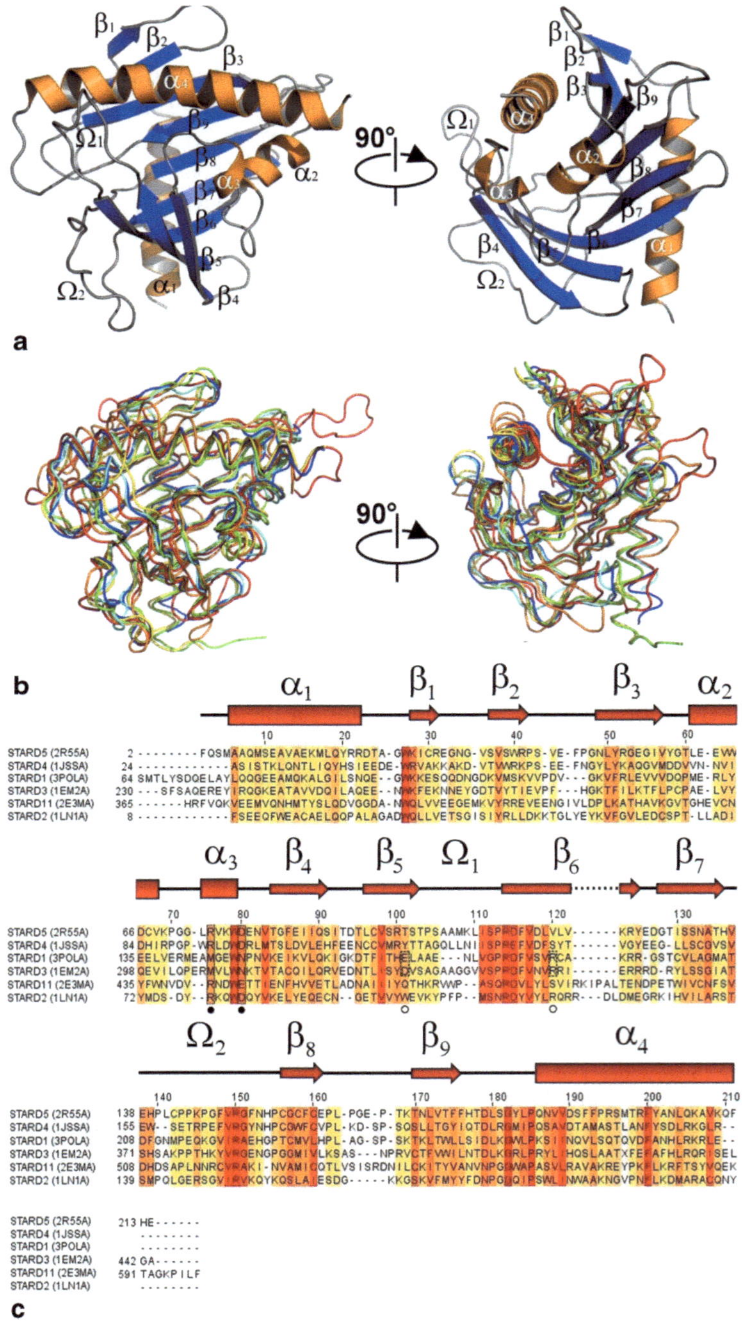

Fig. 3.1 Structure of the START domain. **a** X-ray structure of STARD5 (PDB: 2R55). Some missing residues (Ala[71], Val[72], Gly[165], and E[166]) in the X-ray structure have been modeled as described [44]. **b** Stereo view of the superposition of the backbone traces from crystal structures of six START

were reviewed elsewhere [23, 24]. In this chapter, we provide a brief description of the six mammalian subfamilies followed by a structural perspective for ligand-binding specificity.

STARD1 Subfamily

This subfamily constitutes the membrane-targeted member of cholesterol- and oxysterol-binding START proteins. The archetypical START domain STARD1 (StAR) is essential for cholesterol transfer into mitochondria of steroidogenic tissues and possesses a mitochondrial signal sequence [10, 11, 25–34]. Mutations in the STARD1 gene can result in a lethal disease called congenital lipoid adrenal hyperplasia (lipoidCAH; [11, 25, 26, 28, 31]).

The second STARD1 subfamily member, STARD3 (metastatic lymph node 64, MLN64), binds cholesterol and is targeted to membranes of the late endosomes by an N-terminal domain [16]. STARD1 and STARD3 cellular localization is evocative of their different roles in cellular cholesterol trafficking [35].

STARD4 Subfamily

STARD4/STARD5/STARD6, members of STARD4 subfamily are closely related to the STARD1/STARD3 subfamily [36]. They are soluble sterol-binding proteins composed essentially of a START domain without specific organelle-targeting sequence.

STARD4 binds cholesterol, 7-hydroperoxycholesterol and 7α-hydroxycholesterol [37, 38]. Expressed in liver and kidneys [36], keratinocytes [39], Kupffer cells, and hepatocytes [40], STARD4 is an important component of cholesterol transport and homeostasis. STARD4 overexpression in mouse hepatocytes leads to an increase in intracellular cholesteryl esters by delivering cholesterol to the esterification enzyme Acyl-CoA: cholesterol acyltransferase-1 [38, 40].

STARD5 is expressed in liver Kupffer cells, kidneys, heart [36], and in immune-related cells (macrophages, monocytes, promyelocytic cells, mast cells, and basophils; [41]). In vitro, STARD5 was reported to bind radiolabeled cholesterol and 25-hydroxycholesterol [36, 38] although interestingly, it does not transfer choles-

---

domains from 3 different subfamilies (*blue*, STARD5 (2R55_A); *cyan*, STARD4 (1JSS_A); *green*, STARD1 (3P0L_A); *yellow*, STARD3 (1EM2_A); *orange*, STARD2 (1LN1_A); *red*, STARD11 (2EM3_A)). **c** Sequence alignment of human STARD1, STARD3, STARD4, STARD5, STARD2, and STARD11 The structure-based sequence alignment of START domains was performed within Pymol (http://www.pymol.org/) with CE alignment [80] and formatted for display in Jalview [81] Names and labels are those available in the Protein Data Bank (PDB) access codes. Secondary structure elements of STARD5 are numbered and shown on *top Black circles* (•) indicate position of the salt-bridge in α3-helix and *open circles* (○) specify the position of the salt-bridge in STARD1 and STARD3 *START* steroidogenic-acute-regulatory-protein-related lipid transfer, *CE* combinatorial extension

terol to mitochondria [42]. However, we recently demonstrated that STARD5 is a bile acid-binding domain rather than a cholesterol-binding domain [43, 44]. Subcellular localization data point to a broad cellular distribution, e.g., cytoplasm, plasma membrane (PM), Golgi, endoplasmic reticulum (ER), for STARD5 in macrophages and renal proximal tubules [24, 41, 45, 46].

STARD6, the third member of the STARD4 subfamily, also binds cholesterol [47, 48]. This START domain protein was first detected in testis germinal cells, suggesting a role during germ cell maturation [36, 49–51]. STARD6 is also expressed in Purkinje cells of the nervous system [49, 52, 53]; it has been hypothesized that STARD6 in conjunction with neurosteroids can participate in the development of these cells [54, 55].

STARD2/PCTP Subfamily

This subfamily is composed of phospholipid/sphingolipid-binding START proteins. STARD2, STARD7, and STARD10 bind phosphatidylcholine (PC) and STARD11 binds ceramide. STARD2 specifically transfers/exchanges PC from the ER to the PM. The crystal structure of STARD2 bound to PC shows the classical α-helix/β-grip fold defining a large hydrophobic cavity. The binding specificity is provided by the choline head group and the binding cavity can accommodate PC with saturated or unsaturated acyl groups of different lengths [56].

STARD7, synthesized as a larger precursor protein and processed to a mature protein, can undergo posttranslational modifications that modulate its functionality [57]. STARD7 has a role in modulating trophoblast cell proliferation, migration, and differentiation [58]. STARD7 extracts PC from the cytoplasmic surfaces of the ER, Golgi apparatus, or PM and delivers it to mitochondria [59].

The third member of this family, STARD10, is highly expressed in the liver and is thought to be involved in regulating bile acid metabolism through the modulation of PPAR α-mediated mechanisms [60]. STARD10 selectively transfers PC and phosphatidylethanolamine and preferentially selects phospholipid species containing a palmitoyl or stearoyl fatty acyl chain at the *sn*-1 position and an unsaturated fatty acyl chain (18:1 or 18:2) at the *sn*-2 position [61].

STARD11, commonly known as ceramide transfer protein (CERT), contains additional motifs for cellular localization and is responsible for the movement of ceramides from the ER to the Golgi membrane [62, 63]. The crystal structure of the STARD11 confirms the α-helix/β-grip fold structure for ceramide binding and supports a mechanism for membrane interaction and ceramide extraction/absorption [64, 65]. STARD11 has strict substrate specificity for ceramides with flexibility for their natural isoforms [66]. Polar groups of ceramides participate in a hydrogen-bonding network with specific amino acid residues in the STARD11 cavity, and depending on the molecular ceramide species, different sets of amino acid residues contribute to the hydrophobic interactions with the ceramide hydrocarbon chains [65].

STARD8/STARD12/STARD13: The Sterile α Motif (SAM)-RhoGAP-START Subfamily

Commonly referred to as the deleted in liver cancer (DLC) family of proteins, this subfamily is characterized by a multi-domain structure that consists of an N-terminal sterile α motif (SAM) followed by a serine-rich region, a RhoGAP domain, and a C-terminal START domain (reviewed in [24]). The functions associated with this subfamily have been attributed mainly to the RhoGAP domain and the role of the START domain is not known. These proteins have tumor suppressor activities [67–69] and focal adhesion localization [70, 71]. The crystal structure for STARD13 (DLC-2) indicates the presence of a smaller binding pocket containing polar residues, making it different from the cholesterol and phospholipid START proteins cavities. The authors propose that a charged lipid would be a likely ligand-binding candidate [17].

STARD14/STARD15: The acyl-CoA Thioesterase Subfamily

The acyl-coenzyme A thioesterase (ACOT) subfamily members STARD14 and STARD15 are also multi-domain START proteins. The ACOT family members hydrolyze the thioester bond of fatty acyl-CoAs to generate free fatty acids and coenzyme A [72]. ACOT11 (STARD14) and ACOT12 (STARD15) are distinctive as they contain C-terminal START domains in addition to the tandem N-terminal hot dog-fold domains [72–74]. The STARD14 crystal structure shows that the C-terminal α-helix is broken into two shorter helices and a unique N-terminal helix (α0) acts as a linker to the thioesterase domain [17]. Electron density reminiscent of a fatty acid filled the binding cavity in the crystal, but the actual ligand is unknown.

STARD9: A Kinesin-3 Family Member

STARD9 is another multi-domain START protein belonging to the kinesin-3 family [75] that is implicated in transporting vesicles and organelles [76–78]. STARD9 has a C-terminal START domain predicted to bind phospholipids and/or sterols [23] although its actual ligand and functions are still unknown.

Many aspects of the structure-function of START domains remain to be explored. Namely, the structural determinants for ligand selectivity and affinity are not well defined and the mechanism through which START domains allow for the entry and exit of ligands also remain elusive. In this chapter we will provide an update on those issues from the existing data in the literature.

Mammalian START Domain Structures

The crystal structures of eight mammalian START domains have been solved so far: STARD1 [17], STARD2 [56], STARD3 [16], STARD4 [79], STARD5 [17], STARD11 [65], STARD13, and STARD15 [17]. STARD11 is the only START domain for which the apo- and ligand-bound structures are known. Of all the START domain proteins reported to bind sterols, only STARD6 remains to be solved. We have recently completed the chemical shift assignments (Biological Magnetic Resonance Data Bank (BMRB) 18777) of the START domain of STARD6 and are in the process of refining its solution nuclear magnetic resonance (NMR) structure. The chemical shifts and initial structure demonstrate that the START domain of STARD6 also adopts the conserved α-helix/β-grip fold. Apart from deletions and insertions in loops, the structures of all the START domains are very similar; we show in Fig. 3.1b the superposition of the backbone of STARD1, STARD2, STARD3, STARD4, STARD5, and STARD11. The backbone of these START domains superimposes onto each other with pairwise root mean square deviations (RMSDs) less than 3.0 Å. Whereas the structures of the START domains of STARD2 and STARD11 have been solved with lipidic ligands (e.g., PC (STARD2), ceramides and synthetic analogs ((1R,3R)-N-(3-Hydroxy-1-hydroxymethyl-3-phenylpropyl)alkanamide (HPA); STARD11)), no structure of a START domain-sterol complex is currently available. Hence, the determinants of the specificity of STARD1, STARD3, STARD4, STARD5, and STARD6 towards cholesterol or other sterols remain to be unravelled.

The Determinants of Ligand-Binding Specificity of START Domains

Subsequent to the first structure determination for a START domain, it was suggested that the shape of the cavity had to play an important role in ligand specificity. To illustrate this, we present the molecular surface of the STARD1 cavity containing one cholesterol molecule (Fig. 3.2a). As can be seen, the shape of the cavity matches perfectly the cholesterol molecule. This implies that limited structural changes are expected upon ligand binding. Indeed, in STARD11, very little changes in side-chain conformations were observed between the apo- and ceramide-bound cavities: besides a reorientation of Trp 473 in the Ω1 loop in the (1R,3R)-N-(3-Hydroxy-1-hydroxymethyl-3-phenylpropyl) alkanamide (HPA)-bound structure, the side-chain conformation, and the cavities are very similar [64]. Interestingly, mutations of Trp 473 and Trp 562 near the C-terminal helix reduce membrane binding and ceramide transport in in vitro assays [64].

Models of cholesterol complexes for STARD1 [15, 18, 82], STARD3 [16, 18], and STARD5 [17] have been proposed from molecular docking and molecular dynamics simulations. In STARD1 and STARD3 models, the presence of a conserved salt-bridge at the bottom of the cavity between an acidic residue (Glu169/STARD1 and Asp332/STARD3) and a basic residue (Arg188/STARD1 and Arg351/STARD3) was proposed to be a key determinant for cholesterol binding [15, 16].

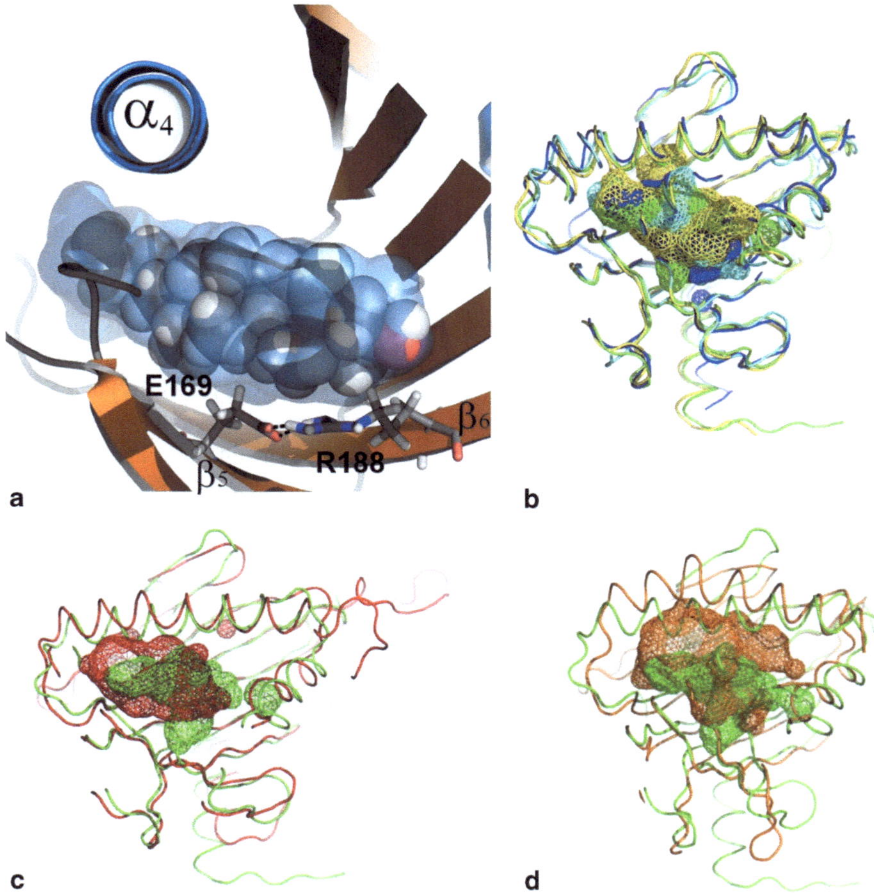

Fig. 3.2 Binding site of the START domain. **a** Model of STARD1in complex with cholesterol (PDB: 2I93; [18]). The cavity surface is displayed and the conserved salt-bridges at the bottom of the cavity between Glu169 and Arg188 are shown. **b** Cavity comparison of STARD1 and STARD4 subfamily members. **c** Comparison between cavities of STARD1 and STARD11. **d** Comparison between cavities of STARD1 and STARD2 Colors and PDB files are the same as in Fig. 3.1. The figure was created with Pymol (http://www.pymol.org/) START steroidogenic-acute-regulatory-protein-related lipid transfer

More precisely, the C3-hydroxyl group of cholesterol could form a specific interaction with the guanidino group of the conserved arginine (Fig. 3.2a). This salt-bridge is not conserved in the STARD4 subfamily (Fig. 3.1c). It has also been suggested that the cholesterol C3-hydroxyl group (OH) could alternately and/or concomitantly interact with the backbone carboxylic group (CO) of Leu199 in STARD1 and the side-chain OH of Ser362, directly or via a structural water molecule [18]. Similarly, Thorsell et al., have presented a model of STARD5-cholesterol complex in which the cholesterol OH interacts with the side-chain OH of Ser132 and the backbone CO of Val68 [17]. Besides the salt-bridge in STARD1 and STARD3, all other proposed binding determinants are located at the bottom of the cavity and superimpose in the primary and tertiary structures. In the model by Thorsell et al., the OH of cholesterol is in close proximity to an arginine-guanidino group that is part of a salt-bridge in helix α3-(Fig. 3.1c; [17]). Notably, this ion pair is structurally conserved in the α3-helix of other mammalian START domain proteins, except for STARD1, STARD3, and STARD9.

It should be noted that the residues forming the cavity are by no means strictly hydrophobic; many side chains with H-bond donors and acceptors line the cavity. This is exemplified in the structure of STARD11 in complex with ceramides; the polar head group of the ceramide within STARD11 is located at the bottom of the cavity where the cholesterol OH is proposed to be buried in STARD1 and STARD3 [15, 16, 18]. More precisely, R442 (α3), E446 (α3), Q467 (β5), N504 (β7) and Y553 (β9) are involved in the stabilization of the polar head-groups through H-bonds [64]. Comparable interactions involving structurally conserved ion pairs and the polar head group of PC are observed in STARD2 [56]. As discussed by Kudo, although ceramides and PC have somewhat similar polar head groups, the STARD11 cavity is too small to accommodate a PC molecule [64]. This reinforces the hypothesis that the ligand specificity displayed by START domains relies on the shape of the cavity and on the formation of polar interactions with the side chains inside the cavity. To further validate this hypothesis, we compared the shape and size of the cavity of the apo-form of STARD11 with those of STARD1, STARD3, STARD4, and STARD5 that bind sterols. As shown in Fig. 3.2b–d, the shape and position of the cavities are different; the cavities in STARD1, STARD3, STARD4, and STARD5 expand deeper towards the bottom. Upon inspection of the side chains present in the bottom depression, highly conserved proline, phenylalanine, and arginine residues as well as the presence of relatively small side chains are observed. Hence, START domains involved in sterol binding have conserved residues that play a structural role in the formation of a cavity that fits the shape of sterol molecules. For STARD1 and STARD4 subfamily members, the structural determinants of sterol binding and specificity point to an intricate interplay between surface complementarity (van der Wall interactions) and H-bonds.

STARD1, STARD5, and STARD6 Ligand Binding Monitored by NMR

Using NMR, we have recently identified residues in STARD5, STARD6, and STARD1 that are involved in their interaction with sterols. More precisely, we used the heteronuclear single quantum coherence (^{1}H-^{15}N-HSQC) experiment to monitor the binding of cholesterol to STARD1 and STARD6 and bile acids to STARD5 [48]. ^{1}H-^{15}N-HSQC are 2D heteronuclear NMR spectra where the correlation between protons and nitrogen nuclei in the backbone and side chain NH and NH2 moieties are recorded. The position (chemical shift) of the correlations (cross-peaks) depends on the local magnetic environment provided by the structure of a protein or complex. Generally, residues that participate in intermolecular interactions will experience a change in their local environment due to the presence of a ligand. Hence, if there is an interaction between a START domain and a ligand inside the cavity, the position of the cross-peaks of the residues involved in the interaction will be perturbed on the ^{1}H-^{15}N-HSQC spectrum. Such residues can be identified unambiguously if the sequence-specific assignments of the chemical shifts of the backbone atoms are available. Using such an approach, it is possible to identify the binding site of a ligand on a protein through titration. In this regard, we have recently completed these assignments for STARD5 [83] and STARD6 (BMRB entry 18777). Finally, long-range (or allosteric) effects can also induce chemical shift changes away from and within the binding site [84].

We have used "soluble cholesterol" in complex with methyl-β-cyclodextrine (β-CD) to titrate cholesterol in STARD1, STARD5, and STARD6. The ^{1}H-^{15}N-HSQC of STARD1 and STARD6 are perturbed when titrated with cholesterol (Fig. 3.3a and b; [48, 85]). Binding was also demonstrated by circular dichroism, which showed an increase of thermodynamic stability of the START domain of STARD1 [86] and STARD6 [47] upon cholesterol addition. However, under similar conditions, cholesterol addition to STARD5 did not induce any apparent changes either in the ^{1}H-^{15}N-HSQC (Fig. 3.3c) or in its temperature denaturation curve [44]. Such results were unexpected since STARD5 has been reported to bind cholesterol [37, 38].

Modeling of the Complex Between STARD5 Primary and Secondary Bile Acids

Given that STARD5 did not bind cholesterol, we investigated more soluble ligands such as bile acids. The rationale of our choice was motivated by the fact that STARD5 is expressed in tissues and cells that are exposed to bile acids. We found that STARD5 specifically binds primary and secondary bile acids [43, 44]. The chemical shift perturbations of STARD5 residues following chenodeoxycholic acid (CDCA) titrations are shown in Fig. 3.3d. These residues circumscribe the binding site of STARD5 and only a few residues with significant chemical shift displacements (CSD) are located away from the cavity, i.e., in β1-, β2- and β3-strands (Fig. 3.4a and b). This indicates that the conformational changes upon binding are minimal and most likely limited to the side chains involved in the binding. Using molecular docking in conjunction with CSD data, we have modeled the interactions between STARD5 and CDCA (Fig. 3.4c; [44]). In the selected poses, all the hydroxyls of the ligand were involved in H-bonds with residues inside the cavity. Moreover, the ligand carboxylate was surrounded by amino acid side chains such as Thr, Ser, and Arg, which can contribute to their stabilization through H-bonds (Fig. 3.4c). The residues most perturbed by the presence of ligands are located around Arg76 (α3), Asp80 (β4), Ser132 (β7), Cys158 (β9), Thr178 and Ser196 (Fig. 3.4c). It is interesting to note that similarly to the STARD5-cholesterol complex model [17],

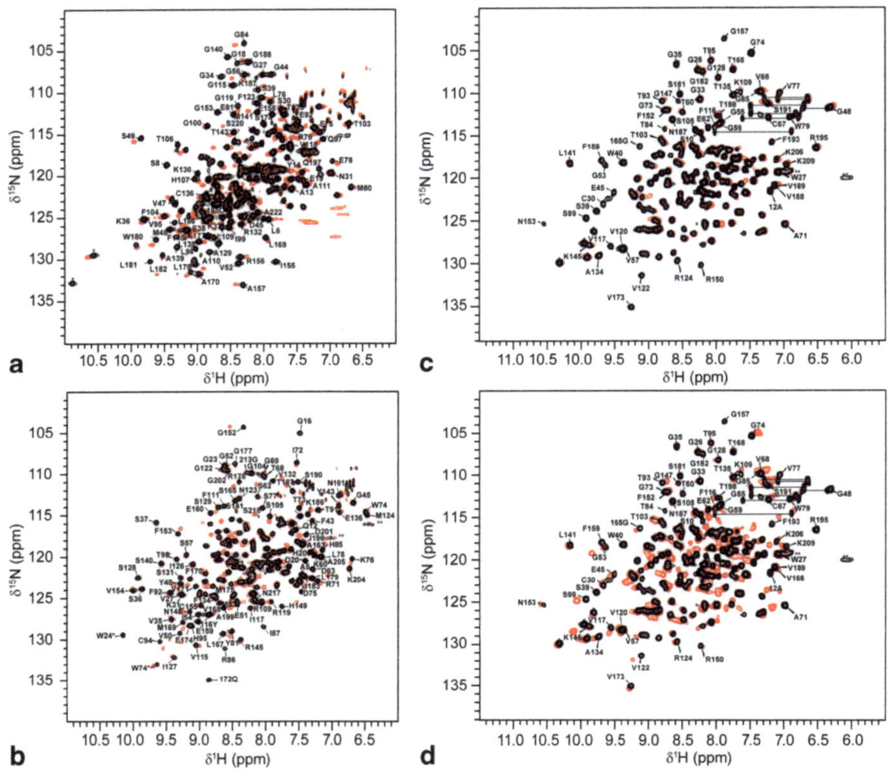

Fig. 3.3 Sterol binding to START domain. ^{1}H-^{15}N-HSQC spectra of STARD1 (**a**), STARD6 (**b**), and STARD5 (**c**) in presence of cholesterol. The human recombinant STARD1 (N-62 StAR), STARD5, and STARD6 with a 6x-histidine tag at the carboxy-terminus was labeled with ^{15}N and purified (*black*). Proteins were incubated with soluble cholesterol (cholesterol complexed with methyl-β-cyclodextrin; *red*) at a 2:1 ratio of cholesterol:protein. Excluding STARD5, several peaks show chemical shift displacements upon cholesterol addition (*red*). Assigned amino acid residues are numbered in each panel. **d** ^{1}H-^{15}N-HSQC spectrum of STARD5 in presence of CDCA at 2:1 ratio of CDCA:protein (*Reprinted from Letourneau D, Lefebvre A, Lavigne P, LeHoux JG. STARD5 specific ligand binding: comparison with STARD1 and STARD4 subfamilies. Mol Cell Endocrinol. 2013 May 22;371(1–2):20–5, with permission from Elsevier*) *START* steroidogenic-acute-regulatory-protein-related lipid transfer, *HSQC* heteronuclear single quantum coherence, *CDCA* chenodeoxycholic acid

H-bonds between the C3-hydroxyl of CDCA and the γ-hydroxyl of Ser132 and the carboxyl of Val68 are observed in our STARD5-bile acid complex. Noteworthy, the hydrophilic face (containing hydroxyl groups) of CDCA is projected towards the salt-bridge between Arg76 (α3) and Asp80 (β4), suggesting an important role for interaction with the polar parts of ligands for the STARD2 and STARD4 subfamilies. Moreover, the C24-carboxylate accepts H-bonds from Thr103 and Thr178 γ1-hydroxyl group leading to the internal solvation of this ionizable moiety. Actually, the intra-molecular salt-bridge between Arg76 and Asp80 could be responsible for the observed CSDs of adjacent residues (Glu81, Asn82, Val83, and Thr84). The

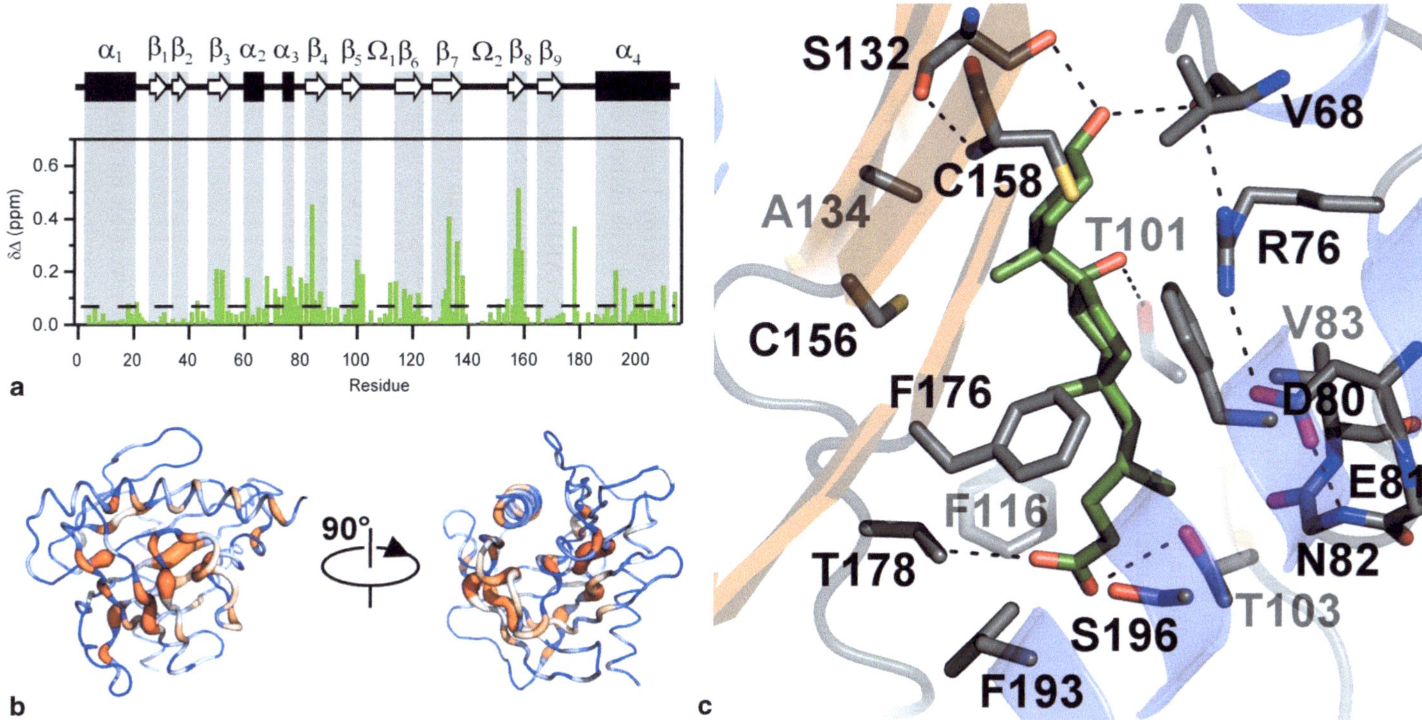

Fig. 3.4 Binding site of STARD5. **a** Weighted CSD (chemical shift displacement) map of binding of CDCA to STARD5. Dashed line represents the mean value of CSD. Secondary structure elements of STARD5 are numbered and shown on top. **b** Mapping of CSD upon CDCA binding. Worm representation with worm radius proportional to the CSD values is coded in a *blue*-to-*orange* gradient; regions unaffected (*blue*) have a thinner backbone worm, regions perturbed (*orange*) are thicker and intermediate regions are white. **c** Representative and corresponding CDCA poses (*green*) found to fulfill (in accordance with the CSD data) optimal H-Bond in STARD5 binding site. The H-bonds suggested that lead to specific binding are highlighted with dashed lines *CSD* chemical shift displacements, *CDCA* chenodeoxycholic acid

specific binding of bile acids to STARD5 has been further validated by isothermal titration calorimetry (ITC) and circular dichroism (CD; [43, 44]). For a more detailed discussion of the potential biological relevance of the interaction between STARD5 and bile acids see the following references [43, 44].

Overall, the existing structural data does not allow straightforward identification of specific features for ligand selectivity. In this regard, it appears that residues in the binding site of STARD1, STARD2, and STARD4 subfamilies define the shape of the cavity, rather than being only a specific set of spatially positioned H-bond donors and acceptors. This indicates that a START domain selectivity for ligands is based predominantly on structural features such as the volume and the shape of ligands.

Ligand Binding and Dissociation Mechanisms

As discussed above, the START domains reported to bind sterols have an internal cavity with a shape that can accommodate a ligand. In this section, we focus on the binding mechanism (entry and exit) of sterols by START domains from structural, dynamical, and thermodynamical point of views.

Cholesterol at the end of the Tunnel

As the structure of the apo-form of STARD3 was first solved by Tsujishita and Hurley [16], the presence of a tunnel-like cavity was unveiled. The authors emphasized the fact that START domains were perfectly designed to bind (or solubilize) single lipid molecules. Based on the fact that STARD1 and STARD3 belong to the same subfamily, they have proposed the existence of such a tunnel in the START domain of STARD1. They also suggested a shuttling mechanism of cholesterol molecules between the outer and inner membrane of mitochondria through that tunnel. This mechanism has since been contested; indeed, when hydrogen atoms (absent from the crystallographic data) are added to STARD1 and STARD3 Protein Data Bank (PDB) files, then more confined cavities are calculated. However, Tsujishita and Hurley also hypothesized the necessity of conformational changes involving the C-terminal helix and/or the Ω1 loop for delivery of sterol molecules [16].

The Intermediate State Model

In 2002, we proposed an alternative mechanism for the reversible binding of cholesterol by N62-STARD1, the STARD1 protein lacking the amino terminal 62 amino acids [15]. Based on the fact that internal cavities destabilize tertiary structures and that the C-terminal helix could move independently from the rest of the molecule, we hypothesized the existence of an intermediate state with the N-terminus intact and the C-terminal helix undergoing a small and reversible local unfolding that is independent of global unfolding [15, 86, 87]. In this intermediate state, the cholesterol-binding site would become accessible and facilitate cholesterol recognition and dissociation. When a cholesterol molecule is in the binding site, the refolding of the C-terminal helix leads to a stable protein; with a lifetime long enough to carry and deliver cholesterol to its target organelle membrane and/or transporter (*see* [88]).

Experimental Validation of the Intermediate State Model

Over the past decade, many studies have provided support for the intermediate state model for STARD1 [15, 18, 32, 33, 86, 87, 89, 90]. Indeed, restricting the movement of the C-terminal of N62-STARD1 hampers its cholesterol-binding affinity and its ability to transfer cholesterol inside mitochondria as determined by pregnenolone production in vitro. Covalently linking helix 4 to the loop between β-strands 1 and 2 with disulfide bridges (S100C/S261C and D106C/A268C) resulted in a decrease of STARD1 cholesterol binding and steroidogenic activity [89]. Adding a reducing agent, however, restored the cholesterol-binding capacity and function of the protein.

CD Spectropolarimetry

The existence of an intermediate state in equilibrium with the native state of N62-STARD1 was further demonstrated by CD. As described in Roostaee et al., the far-UV CD spectrum of N62-STARD1 with an equimolar concentration of cholesterol induces more negative molar ellipticities than control, indicative of an increase in secondary structure (α and/or β; [86]). Moreover, the presence of an isosbestic point at 203 nm strongly suggests that cholesterol addition stabilizes the C-terminus into an α-helix configuration rather than a random coil structure. In addition, the melting temperature of the cholesterol-N62-STARD1 (1:1) complex is increased compared to the free construct [86]. Finally, by monitoring the increase in α-helical content at 222 nm, we have titrated N62-STARD1 with cholesterol and confirmed that STARD1 binds cholesterol in a 1:1 stoichiometry with apparent affinity of $\sim 3 \times 10^{-8}$ M [86].

Solution-state NMR

In our effort to solve the solution structure of N62-STARD1 we are in the process of completing the assignments of its backbone chemical shifts (^{1}H, ^{13}C, and ^{15}N). Up to now, we have been able to assign 70 % of the resonances. Most of the missing assignments are located in α4-helix. The difficulty resides in the fact that the resonances are either missing, overlapped, or broadened. This is representative of a helix that is either unfolded and/or undergoing ms-ms conformational exchange. While this is not a bona fide proof of the existence of the intermediate state, it strongly suggests that α4-helix is undergoing local unfolding or motions independent of the rest of the protein. The validation of this hypothesis will necessitate the completion of the backbone assignment and the characterization of the backbone molecular dynamic on the ps-ns and ms-ms timescale using spin relaxation methods.

The Molten Globule Hypothesis

In the late 1990s, Walter Miller's group discovered that the N62-STARD1 domain behaved like a molten globule at acidic pH (from pH 4.5 to 3). Whereas the amount of secondary structure did not significantly change from pH 4.5 to 3, the thermodynamic stability at 25 °C (i.e., $\Delta G^{\circ}_{u}(25^{\circ})$) of the N62- STARD1 decreased significantly [90]. Note that the stability of N62-STARD1 as measured by urea denaturation was observed to be constant from pH 8.3 to 4.5. This means that while the secondary structure is maintained as pH is decreased, the amount of tertiary structure is diminished at pH values lower than 4.5. This is in agreement with the existence of a molten globule, i.e., a protein with native-like secondary structure and a loosely packed tertiary structure [91]. In addition, it was found that the C-terminal region (193–285) was less protected from proteolysis at acidic pH than the N-terminus (63–188; [42, 90]). These data lead to the hypothesis that low pH at or near the surface of mitochondria induces a molten globular state of STARD1 with the N-terminus more tightly folded than the C-terminus and that this state is involved in cholesterol binding and transfer to the inner membrane of mitochondria. However, it is hard to reason that such a low pH value (4.5) can be achieved near mitochondria and more importantly, how a molten globule state can specifically bind cholesterol or any other ligand [32].

In order to evaluate the possibility that the molten globular state of N62-STARD1 can bind cholesterol, we have analyzed its thermodynamical stability with and without cholesterol using CD [86]. We reasoned that if cholesterol binds to the molten globule state, then the apparent thermodynamical stability of N62-STARD1 should be increased by the stabilization free energy provided by cholesterol binding. The far-UV CD spectra of N62-STARD1 displayed a sizable content of secondary structure at acidic pH. However, as indicated by the lack of cooperativity on temperature denaturation curves monitored by CD, N62-STARD1 is devoid of stable tertiary structure. In addition, this lack of cooperativity is persistent in the presence of cholesterol. These data indicate that the molten globule state cannot provide the minimal tertiary structure necessary for cholesterol binding.

Ω1-Loop and α3-Helix Gating

In 2006, Murcia et al., proposed another mechanism for STARD1 cholesterol binding and release [18]. After cholesterol docking in the STARD3 structure and in a STARD1 model, these authors proceeded to molecular dynamics simulations for cholesterol exit from the internal cavity. By steering the dissociation of cholesterol from the START domains, they demonstrated that exit was possible through a path created by a conformational change (opening) in the Ω1-loop. It was also assumed that ligand entry would follow the same route. However, although molecular dynamics simulations reported by this group suggest the presence of movements in the Ω-loop, experimental validation is necessary to confirm the presence of such movement for entry and exit paths.

In this regard, Kudo et al., have reported high crystallographic B-factors in the Ω1-loop and to a lesser extent in α3-helix of STARD11 [64]. High B-factors are caused by structural flexibility of the scattering electron shells of the nuclei of the protein in the crystal and are proportional to the amplitude of motions. Although B-factors do not report on the motion time scale, they are real molecular motions. Thus, these data support the presence of motions in the Ω1-loop and a possible role of the loop in gating the ligand entry and exit into the START domain internal cavity. It should be noted that the molecular motions unveiled by B-factors will depend on the crystal packing within and between the unit cells. Based on this observation, Kudo et al., proposed that movements in the Ω1-loop and α3-helix could act as a gate for ligand entry and exit. Irrespective of the time-scale, movements of high amplitude in the Ω1-loop and α3-helix will weaken the tertiary interaction stabilizing the C-terminal helix in its folded conformation.

We have recently monitored the molecular dynamics of the backbone NH chemical shift of the apo-form of STARD5 on the ps-ns time scale by ^{15}N spin relaxation (i.e., heteronuclear $\{^{1}H\}$-^{15}N NOE). Significant movements in the α3-helix and to a lesser extent in the Ω1-loop are noted on this time scale. Interestingly, the location of a more mobile backbone NH is in accordance with higher B-factors location taken from the X-ray analysis of the apo-form of STARD5 (PDB: 2R55). A similar distribution of molecular motions in STARD6 has been recorded (unpublished data). Taken together, experimental data and simulations reviewed in this section support the presence of mobility in Ω1-loop and α3-helix on the ps-ns time scale.

Towards a Unified Model for the Binding and Delivery of Cholesterol and Lipids

Although movements on the ps-ns time scale in a complex between a START domain and a molecule of cholesterol can be expected to lead to the exit of the ligand, it is still puzzling how such fast librations could allow for the entry. It seems logical that larger conformational changes would be needed in order to let the ligand diffuse

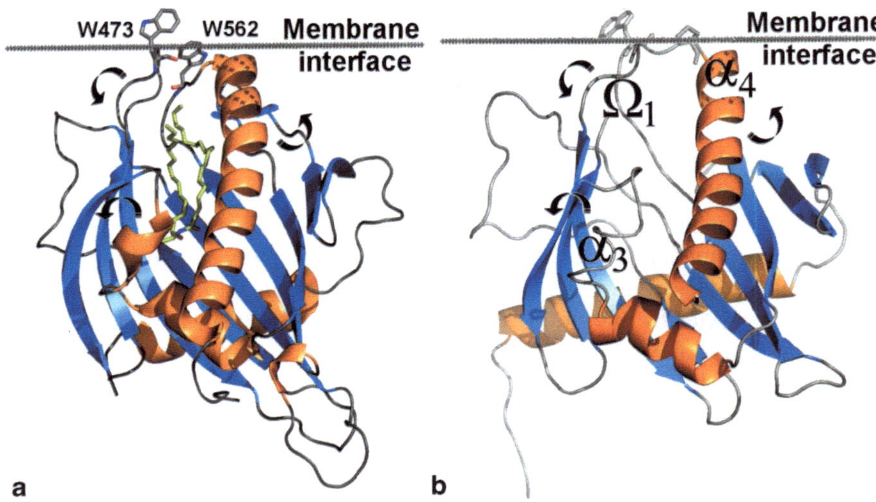

Fig. 3.5 Orientation of START domain at membrane interface. **a** Predicted membrane anchoring mode of STARD11. The two tryptophan residues altering ceramide (*green*) transfer efficiency and its interaction with membranes are shown. **b** STARD1 membrane anchoring mode and potential molecular motion in the Ω1-loop, α3-, and α4-helices (*black arrow*) in the proposed mechanism *START* steroidogenic-acute-regulatory-protein-related lipid transfer

inside the cavity. In this regard, it is possible that the movements on the ps-ns can serve to activate larger conformational changes such as opening of the C-terminal helix. However, lipid molecules in membranes have limited degrees of freedom compared to water soluble ligands; in membrane molecular frames, apart from diffusing in the plane membrane, lipids can only rotate along their Z axis (normal to the membrane plane) with their polar head groups facing water. Once a START domain adsorbs onto a membrane, the diffusion inside the binding site should be facilitated. Kudo et al., demonstrated that mutations of Trp (473 and 562) residues to Ala in the Ω1-loop and near the N-terminus of the α4-helix, respectively, altered the interaction of STARD11 with membranes and decreased ceramide transfer efficiency [64]. Aromatic side chains have a propensity to partition at the membrane-water interface and play a key role into the recruitment of peripheral membrane proteins [92, 93]. Recently, an extensive bioinformatics search has revealed the presence of such anchoring motifs on all the START domains of known structures[1] [94, 95]. In complete agreement with their functional role unveiled by Kudo et al., Trp 473 and 562 of STARD11 are both located on the same side of the protein and are part of this motif (Fig. 3.5a; [64]). Noteworthy, STARD1, STARD2, STARD3, STARD4, and STARD5 display a similar anchoring mode to the membrane[2]. It is

[1] The orientations of proteins in membranes (OPM) database are freely accessible at http://opm.phar.umich.edu.

[2] Orientation of START domain at membrane interface (http://opm.phar.umich.edu/families.php?family=204).

also interesting to note that the N-terminus of the α4-helix is part of the proposed anchoring interface and in an orientation favorably positioned to interact with the membrane surface (Fig. 3.5b). Interestingly, earlier data on STARD1 have shown that the C-terminal helix was protected from proteolysis when small unilamellar vesicles (SUV) were present, clearly indicating an affinity for the C-terminal helix for lipid membranes [82].

In this context, it can be envisioned that conformational changes in the Ω1-loop, α3- and α4-helices stabilized by interactions with the membrane, favored by aromatic residues at precise anchoring points, should allow enough space and time for the diffusion of lipid molecules, polar head first, into the cavity (Fig. 3.5a and b). This proposed mechanism is in agreement with the cross-linking result of STARD1 α4-helix where the cholesterol binding and pregnenolone production in vitro was reduced [89].

Summary

The resolution of the detailed mechanism of ligand binding (and delivery) from (and to) membranes is a very challenging task. It will necessitate the thermodynamical and structural characterization of the interaction of apo-START domains with membranes and the detailed measurements of the kinetics of ligand entry and release using a combination of biophysical techniques and molecular dynamics simulations. With regard to ligand-selectivity determinants, the resolution of the structures of STARD1 and STARD4 subfamily members in complex with sterols is of utmost importance. In addition, structure determination of mutants will be necessary to validate the importance of the amino acid side chains that define the shape of the binding site versus their role in specific interactions with the polar moieties of sterols. This is particularly important to understand the role of mutations in the STARD1 gene in the structural etiology of lipoidCAH.

Acknowledgements This work was supported by a grant (MT–10983) to JGL and PL from the Canadian Institutes of Health Research and by the Regroupement Stratégique sur la Structure, la Fonction et l'Ingénierie des Protéines (PROTEO).

References

1. Lehoux JG, Sandor T, Henderson IW, Jones IC. Some aspects of the dietary sodium intake on the regulation of aldosterone biosynthesis in rat adrenals. Can J Biochem. 1974 Jan;52(1):1–6.
2. LeHoux J-G, Bernard H, Ducharme L, Lefebvre A, Shapcott D, Tremblay A, et al. The regulation of the formation of glucocorticoids and mineralocorticoids in vivo. In: Bittar EE, editor. Advances in molecular and cell biology. Elsevier: London; 1996. pp. 149–201.
3. Karaboyas GC, Koritz SB. Identity of the site of action of 3′,5′-adenosine monophosphate and adrenocortocotropic hormone in corticosteroidogenesis in rat adrenal and beef adrenal cortex slices*. Biochemistry. 1965 March 1;4(3):462–8.

4. Stone D, Hechter O. Studies on ACTH action in perfused bovine adrenals: the site of action of ACTH in corticosteroidogenesis. Arch Biochem Biophys. 1954 Aug;51(2):457–69.
5. Privalle CT, Crivello JF, Jefcoate CR. Regulation of intramitochondrial cholesterol transfer to side-chain cleavage cytochrome P-450 in rat adrenal gland. Proc Natl Acad Sci U S A. 1983 Feb;80(3):702–6.
6. Yago N, Ichii S. Submitochondrial distribution of components of the steroid 11 beta-hydroxylase and cholesterol sidechain-cleaving enzyme systems in hog adrenal cortex. J Biochem. 1969 Feb;65(2):215–24.
7. Arthur JR, Boyd GS. The effect of inhibitors of protein synthesis on cholesterol side-chain cleavage in the mitochondria of luteinized rat ovaries. Eur J Biochem. 1974 Nov 1;49(1):117–27.
8. Epstein LF, Orme-Johnson NR. Regulation of steroid hormone biosynthesis. Identification of precursors of a phosphoprotein targeted to the mitochondrion in stimulated rat adrenal cortex cells. J Biol Chem. 1991 Oct 15;266(29):19739–45.
9. Krueger RJ, Orme-Johnson NR. Acute adrenocorticotropic hormone stimulation of adrenal corticosteroidogenesis. Discovery of a rapidly induced protein. J Biol Chem. 1983 Aug 25;258(16):10159–67.
10. Clark BJ, Wells J, King SR, Stocco DM. The purification, cloning, and expression of a novel luteinizing hormone-induced mitochondrial protein in MA-10 mouse Leydig tumor cells. Characterization of the steroidogenic acute regulatory protein (StAR). J Biol Chem. 1994 Nov 11;269(45):28314–22.
11. Lin D, Sugawara T, Strauss JF, 3rd, Clark BJ, Stocco DM, Saenger P, et al. Role of steroidogenic acute regulatory protein in adrenal and gonadal steroidogenesis. Science. 1995 March 24;267(5205):1828–31.
12. Fleury A, Ducharme L, LeHoux JG. In vivo effects of adrenocorticotrophin on the expression of the hamster steroidogenic acute regulatory protein. J Mol Endocrinol. 1998 Oct;21(2): 131–9.
13. Lehoux JG, Hales DB, Fleury A, Briere N, Martel D, Ducharme L. The in vivo effects of adrenocorticotropin and sodium restriction on the formation of the different species of steroidogenic acute regulatory protein in rat adrenal. Endocrinology. 1999 Nov;140(11):5154–64.
14. Fleury A, Mathieu AP, Ducharme L, Hales DB, LeHoux JG. Phosphorylation and function of the hamster adrenal steroidogenic acute regulatory protein (StAR). J Steroid Biochem Mol Biol. 2004 Aug;91(4–5):259–71.
15. Mathieu AP, Fleury A, Ducharme L, Lavigne P, LeHoux JG. Insights into steroidogenic acute regulatory protein (StAR)-dependent cholesterol transfer in mitochondria: evidence from molecular modeling and structure-based thermodynamics supporting the existence of partially unfolded states of StAR. J Mol Endocrinol. 2002 Dec;29(3):327–45.
16. Tsujishita Y, Hurley JH. Structure and lipid transport mechanism of a StAR-related domain. Nat Struct Biol. 2000 May;7(5):408–14.
17. Thorsell AG, Lee WH, Persson C, Siponen MI, Nilsson M, Busam RD, et al. Comparative structural analysis of lipid binding START domains. PLoS One. 2011;6(6):e19521.
18. Murcia M, Faraldo-Gomez JD, Maxfield FR, Roux B. Modeling the structure of the StART domains of MLN64 and StAR proteins in complex with cholesterol. J Lipid Res. 2006 Dec;47(12):2614–30.
19. Iyer LM, Koonin EV, Aravind L. Adaptations of the helix-grip fold for ligand binding and catalysis in the START domain superfamily. Proteins. 2001 May 1;43(2):134–44.
20. Schrick K, Nguyen D, Karlowski WM, Mayer KF. START lipid/sterol-binding domains are amplified in plants and are predominantly associated with homeodomain transcription factors. Genome Biol. 2004;5(6):R41.
21. Mathieu AP, Lavigne P, LeHoux JG. Molecular modeling and structure-based thermodynamic analysis of the StAR protein. Endocr Res. 2002 Nov;28(4):419–23.
22. Ponting CP, Aravind L. START: a lipid-binding domain in StAR, HD-ZIP and signalling proteins. Trends Biochem Sci. 1999 April;24(4):130–2.

23. Alpy F, Tomasetto C. Give lipids a START: the StAR-related lipid transfer (START) domain in mammals. J Cell Sci. 2005 July 1;118(Pt 13):2791–801.
24. Clark BJ. The mammalian START domain protein family in lipid transport in health and disease. J Endocrinol. 2012 March;212(3):257–75.
25. Bhangoo A, Gu WX, Pavlakis S, Anhalt H, Heier L, Ten S, et al. Phenotypic features associated with mutations in steroidogenic acute regulatory protein. J Clin Endocrinol Metab. 2005 Nov;90(11):6303–9.
26. Bose HS, Sugawara T, Strauss JF, 3rd, Miller WL. The pathophysiology and genetics of congenital lipoid adrenal hyperplasia. N Engl J Med. 1996 Dec 19;335(25):1870–8.
27. Chen X, Baker BY, Abduljabbar MA, Miller WL. A genetic isolate of congenital lipoid adrenal hyperplasia with atypical clinical findings. J Clin Endocrinol Metab. 2005 Feb;90(2):835–40.
28. Fluck CE, Pandey AV, Dick B, Camats N, Fernandez-Cancio M, Clemente M, et al. Characterization of novel StAR (steroidogenic acute regulatory protein) mutations causing non-classic lipoid adrenal hyperplasia. PLoS One. 2011;6(5):e20178.
29. Gassner HL, Toppari J, Quinteiro Gonzalez S, Miller WL. Near-miss apparent SIDS from adrenal crisis. J Pediatr. 2004 Aug;145(2):178–83.
30. Katsumata N, Kawada Y, Yamamoto Y, Noda M, Nimura A, Horikawa R, et al. A novel compound heterozygous mutation in the steroidogenic acute regulatory protein gene in a patient with congenital lipoid adrenal hyperplasia. J Clin Endocrinol Metab. 1999 Nov;84(11): 3983–7.
31. Khoury J, Amundsen AL, Tonstad S, Henriksen T, Ose L, Retterstol K, et al. Evidence for impaired physiological decrease in the uteroplacental vascular resistance in pregnant women with familial hypercholesterolemia. Acta Obstet Gynecol Scand. 2009;88(2):222–6.
32. Lavigne P, Najmanivich R, Lehoux JG. Mammalian StAR-related lipid transfer (START) domains with specificity for cholesterol: structural conservation and mechanism of reversible binding. Subcell Biochem. 2010;51:425–37.
33. Lehoux JG, Mathieu A, Lavigne P, Fleury A. Adrenocorticotropin regulation of steroidogenic acute regulatory protein. Microsc Res Tech. 2003 June 15;61(3):288–99.
34. Nakae J, Tajima T, Sugawara T, Arakane F, Hanaki K, Hotsubo T, et al. Analysis of the steroidogenic acute regulatory protein (StAR) gene in Japanese patients with congenital lipoid adrenal hyperplasia. Hum Mol Genet. 1997 April;6(4):571–6.
35. Alpy F, Tomasetto C. MLN64 and MENTHO, two mediators of endosomal cholesterol transport. Biochem Soc Trans. 2006 June;34(Pt 3):343–5.
36. Soccio RE, Adams RM, Romanowski MJ, Sehayek E, Burley SK, Breslow JL. The cholesterol-regulated StarD4 gene encodes a StAR-related lipid transfer protein with two closely related homologues, StarD5 and StarD6. Proc Natl Acad Sci U S A. 2002 May 14;99(10):6943–8.
37. Rodriguez-Agudo D, Ren S, Hylemon PB, Redford K, Natarajan R, Del Castillo A, et al. Human StarD5, a cytosolic StAR-related lipid binding protein. J Lipid Res. 2005 Aug;46(8):1615–23.
38. Rodriguez-Agudo D, Ren S, Wong E, Marques D, Redford K, Gil G, et al. Intracellular cholesterol transporter StarD4 binds free cholesterol and increases cholesteryl ester formation. J Lipid Res. 2008 July;49(7):1409–19.
39. Elbadawy HM, Borthwick F, Wright C, Martin PE, Graham A. Cytosolic StAR-related lipid transfer domain 4 (STARD4) protein influences keratinocyte lipid phenotype and differentiation status. Br J Dermatol. 2011 March;164(3):628–32.
40. Rodriguez-Agudo D, Calderon-Dominguez M, Ren S, Marques D, Redford K, Medina-Torres MA, et al. Subcellular localization and regulation of StarD4 protein in macrophages and fibroblasts. Biochim Biophys Acta. 2011 Oct;1811(10):597–606.
41. Rodriguez-Agudo D, Ren S, Hylemon PB, Montanez R, Redford K, Natarajan R, et al. Localization of StarD5 cholesterol binding protein. J Lipid Res. 2006 June;47(6):1168–75.
42. Bose HS, Whittal RM, Ran Y, Bose M, Baker BY, Miller WL. StAR-like activity and molten globule behavior of StARD6, a male germ-line protein. Biochemistry. 2008 Feb 26;47(8):2277–88.

43. Letourneau D, Lorin A, Lefebvre A, Cabana J, Lavigne P, Lehoux JG. Thermodynamic and solution state NMR characterization of the binding of secondary and conjugated bile acids to STARD5. Biochim Biophys Acta. 2013 July 16;1831(11):1589–99.
44. Letourneau D, Lorin A, Lefebvre A, Frappier V, Gaudreault F, Najmanovich R, et al. StAR-related lipid transfer domain protein 5 binds primary bile acids. J Lipid Res. 2012 Dec;53(12):2677–89.
45. Chen YC, Meier RK, Zheng S, Khundmiri SJ, Tseng MT, Lederer ED, et al. Steroidogenic acute regulatory-related lipid transfer domain protein 5 localization and regulation in renal tubules. Am J Physiol Renal Physiol. 2009 Aug;297(2):F380–8.
46. Rodriguez-Agudo D, Calderon-Dominguez M, Medina MA, Ren S, Gil G, Pandak WM. ER stress increases StarD5 expression by stabilizing its mRNA and leads to relocalization of its protein from the nucleus to the membranes. J Lipid Res. 2012;53(12):2708–15.
47. Letourneau D, Lefebvre A, Frappier V, LeHoux JG, Lavigne P. Structure, dynamics and cholesterol binding site characterization of STARD6 from solution state NMR studies. Endocr Rev. 2012 June;33, Meeting Abstracts, SUN-234(Meeting Abstracts).
48. Letourneau D, Lefebvre A, Lavigne P, LeHoux JG. STARD5 specific ligand binding: comparison with STARD1 and STARD4 subfamilies. Mol Cell Endocrinol. 2013 May 22;371 (1–2):20–5.
49. Chang IY, Kim JK, Lee SM, Kim JN, Soh J, Kim JW, et al. The changed immunoreactivity of StarD6 after pilocarpine-induced epilepsy. Neuroreport. 2009 July 1;20(10):963–7.
50. Chang IY, Shin SY, Kim JW, Yu JM, Kim JS, Song PI, et al. The changed immunolocalization of START-domain-containing 6 (StarD6) during the development of testes in rat perinatal hypothyroidism. Acta Histochem. 2007;109(4):315–21.
51. Gomes C, Oh SD, Kim JW, Chun SY, Lee K, Kwon HB, et al. Expression of the putative sterol binding protein Stard6 gene is male germ cell specific. Biol Reprod. 2005 March;72(3):651–8.
52. Chang IY, Jeon YJ, Jung SM, Jang YH, Ahn JB, Park KS, et al. Does the StarD6 mark the same as the StAR in the nervous system? J Chem Neuroanat. 2010 Nov;40(3):239–42.
53. Chang IY, Kim JH, Hwang G, Song PI, Song RJ, Kim JW, et al. Immunohistochemical detection of StarD6 in the rat nervous system. Neuroreport. 2007 Oct 8;18(15):1615–9.
54. Chang IY, Ohn T, Ko GS, Yoon Y, Kim JW, Yoon SP. Immunolocalization of steroidogenic acute regulatory protein-related lipid transfer (START) domain-containing proteins in the developing cerebellum of normal and hypothyroid rats. J Chem Neuroanat. 2012 Jan;43(1):28–33.
55. Tsutsui K. Biosynthesis, mode of action and functional significance of neurosteroids in the developing Purkinje cell. J Steroid Biochem Mol Biol. 2006 Dec;102(1–5):187–94.
56. Roderick SL, Chan WW, Agate DS, Olsen LR, Vetting MW, Rajashankar KR, et al. Structure of human phosphatidylcholine transfer protein in complex with its ligand. Nat Struct Biol. 2002 July;9(7):507–11.
57. Flores-Martin J, Rena V, Angeletti S, Panzetta-Dutari GM, Genti-Raimondi S. The lipid transfer protein StarD7: structure, function, and regulation. Int J Mol Sci. 2013;14(3):6170–86.
58. Flores-Martin J, Rena V, Marquez S, Panzetta-Dutari GM, Genti-Raimondi S. StarD7 knockdown modulates ABCG2 expression, cell migration, proliferation, and differentiation of human choriocarcinoma JEG-3 cells. PLoS One. 2012;7(8):e44152.
59. Horibata Y, Sugimoto H. StarD7 mediates the intracellular trafficking of phosphatidylcholine to mitochondria. J Biol Chem. 2010 March 5;285(10):7358–65.
60. Ito M, Yamanashi Y, Toyoda Y, Izumi-Nakaseko H, Oda S, Sugiyama A, et al. Disruption of Stard10 gene alters the PPARalpha-mediated bile acid homeostasis. Biochim Biophys Acta. 2013 Feb;1831(2):459–68.
61. Olayioye MA, Vehring S, Muller P, Herrmann A, Schiller J, Thiele C, et al. StarD10, a START domain protein overexpressed in breast cancer, functions as a phospholipid transfer protein. J Biol Chem. 2005 July 22;280(29):27436–42.

62. Hanada K, Kumagai K, Tomishige N, Yamaji T. CERT-mediated trafficking of ceramide. Biochim Biophys Acta. 2009 July;1791(7):684–91.
63. Hanada K, Kumagai K, Yasuda S, Miura Y, Kawano M, Fukasawa M, et al. Molecular machinery for non-vesicular trafficking of ceramide. Nature. 2003 Dec 18;426(6968):803–9.
64. Kudo N, Kumagai K, Matsubara R, Kobayashi S, Hanada K, Wakatsuki S, et al. Crystal structures of the CERT START domain with inhibitors provide insights into the mechanism of ceramide transfer. J Mol Biol. 2010 Feb 19;396(2):245–51.
65. Kudo N, Kumagai K, Tomishige N, Yamaji T, Wakatsuki S, Nishijima M, et al. Structural basis for specific lipid recognition by CERT responsible for nonvesicular trafficking of ceramide. Proc Natl Acad Sci U S A. 2008 Jan 15;105(2):488–93.
66. Kumagai K, Yasuda S, Okemoto K, Nishijima M, Kobayashi S, Hanada K. CERT mediates intermembrane transfer of various molecular species of ceramides. J Biol Chem. 2005 Feb 25;280(8):6488–95.
67. Ching YP, Wong CM, Chan SF, Leung TH, Ng DC, Jin DY, et al. Deleted in liver cancer (DLC) 2 encodes a RhoGAP protein with growth suppressor function and is underexpressed in hepatocellular carcinoma. J Biol Chem. 2003 March 21;278(12):10824–30.
68. Durkin ME, Ullmannova V, Guan M, Popescu NC. Deleted in liver cancer 3 (DLC-3), a novel Rho GTPase-activating protein, is downregulated in cancer and inhibits tumor cell growth. Oncogene. 2007 July 5;26(31):4580–9.
69. Durkin ME, Yuan BZ, Zhou X, Zimonjic DB, Lowy DR, Thorgeirsson SS, et al. DLC-1: a Rho GTPase-activating protein and tumour suppressor. J Cell Mol Med. 2007 Sept–Oct;11(5):1185–207.
70. Kawai K, Iwamae Y, Yamaga M, Kiyota M, Ishii H, Hirata H, et al. Focal adhesion-localization of START-GAP1/DLC1 is essential for cell motility and morphology. Genes Cells. 2009 Feb;14(2):227–41.
71. Kim TY, Healy KD, Der CJ, Sciaky N, Bang YJ, Juliano RL. Effects of structure of Rho GTPase-activating protein DLC-1 on cell morphology and migration. J Biol Chem. 2008 Nov 21;283(47):32762–70.
72. Kirkby B, Roman N, Kobe B, Kellie S, Forwood JK. Functional and structural properties of mammalian acyl-coenzyme A thioesterases. Prog Lipid Res. 2010 Oct;49(4):366–77.
73. Hunt MC, Yamada J, Maltais LJ, Wright MW, Podesta EJ, Alexson SE. A revised nomenclature for mammalian acyl-CoA thioesterases/hydrolases. J Lipid Res. 2005 Sept;46(9):2029–32.
74. Kang HW, Niepel MW, Han S, Kawano Y, Cohen DE. Thioesterase superfamily member 2/acyl-CoA thioesterase 13 (Them2/Acot13) regulates hepatic lipid and glucose metabolism. FASEB J. 2012 May;26(5):2209–21.
75. Torres JZ, Summers MK, Peterson D, Brauer MJ, Lee J, Senese S, et al. The STARD9/Kif16a kinesin associates with mitotic microtubules and regulates spindle pole assembly. Cell. 2011 Dec 9;147(6):1309–23.
76. Hoepfner S, Severin F, Cabezas A, Habermann B, Runge A, Gillooly D, et al. Modulation of receptor recycling and degradation by the endosomal kinesin KIF16B. Cell. 2005 May 6;121(3):437–50.
77. Miki H, Okada Y, Hirokawa N. Analysis of the kinesin superfamily: insights into structure and function. Trends Cell Biol. 2005 Sept;15(9):467–76.
78. Nangaku M, Sato-Yoshitake R, Okada Y, Noda Y, Takemura R, Yamazaki H, et al. KIF1B, a novel microtubule plus end-directed monomeric motor protein for transport of mitochondria. Cell. 1994 Dec 30;79(7):1209–20.
79. Romanowski MJ, Soccio RE, Breslow JL, Burley SK. Crystal structure of the Mus musculus cholesterol-regulated START protein 4 (StarD4) containing a StAR-related lipid transfer domain. Proc Natl Acad Sci U S A. 2002 May 14;99(10):6949–54.
80. Shindyalov IN, Bourne PE. Protein structure alignment by incremental combinatorial extension (CE) of the optimal path. Protein Eng. 1998 Sept;11(9):739–47.
81. Waterhouse AM, Procter JB, Martin DM, Clamp M, Barton GJ. Jalview Version 2-a multiple sequence alignment editor and analysis workbench. Bioinformatics. 2009 May 1;25(9):1189–91.

82. Yaworsky DC, Baker BY, Bose HS, Best KB, Jensen LB, Bell JD, et al. pH-dependent Interactions of the carboxyl-terminal helix of steroidogenic acute regulatory protein with synthetic membranes. J Biol Chem. 2005 Jan 21;280(3):2045–54.
83. Lorin A, Letourneau D, Lefebvre A, Lehoux JG, Lavigne P. (1)H, (13)C, and (15)N backbone chemical shift assignments of StAR-related lipid transfer domain protein 5 (STARD5). Biomol NMR Assign. 2012 March 6;7(1):21–4.
84. Stevens SY, Sanker S, Kent C, Zuiderweg ER. Delineation of the allosteric mechanism of a cytidylyltransferase exhibiting negative cooperativity. Nat Struct Biol. 2001 Nov;8(11): 947–52.
85. Barbar E, Lehoux JG, Lavigne P. Toward the NMR structure of StAR. Mol Cell Endocrinol. 2009 March 5;300(1–2):89–93.
86. Roostaee A, Barbar E, Lehoux JG, Lavigne P. Cholesterol binding is a prerequisite for the activity of the steroidogenic acute regulatory protein (StAR). Biochem J. 2008 June 15;412(3):553–62.
87. Roostaee A, Barbar E, Lavigne P, LeHoux JG. The mechanism of specific binding of free cholesterol by the steroidogenic acute regulatory protein: evidence for a role of the C-terminal alpha-helix in the gating of the binding site. Biosci Rep. 2009 April;29(2):89–101.
88. Rone MB, Fan J, Papadopoulos V. Cholesterol transport in steroid biosynthesis: role of protein-protein interactions and implications in disease states. Biochim Biophys Acta. 2009 July;1791(7):646–58.
89. Baker BY, Yaworsky DC, Miller WL. A pH-dependent molten globule transition is required for activity of the steroidogenic acute regulatory protein, StAR. J Biol Chem. 2005 Dec 16;280(50):41753–60.
90. Bose HS, Whittal RM, Baldwin MA, Miller WL. The active form of the steroidogenic acute regulatory protein, StAR, appears to be a molten globule. Proc Natl Acad Sci U S A. 1999 June 22;96(13):7250–5.
91. Ptitsyn OB. Molten globule and protein folding. Adv Protein Chem. 1995;47:83–229.
92. White SH, Wimley WC. Hydrophobic interactions of peptides with membrane interfaces. Biochim Biophys Acta. 1998 Nov 10;1376(3):339–52.
93. Wimley WC, White SH. Experimentally determined hydrophobicity scale for proteins at membrane interfaces. Nat Struct Biol. 1996 Oct;3(10):842–8.
94. Lomize AL, Pogozheva ID, Lomize MA, Mosberg HI. The role of hydrophobic interactions in positioning of peripheral proteins in membranes. BMC Struct Biol. 2007;7:44.
95. Lomize MA, Pogozheva ID, Joo H, Mosberg HI, Lomize AL. OPM database and PPM web server: resources for positioning of proteins in membranes. Nucleic Acids Res. 2012 Jan;40(Database issue):D370–6.

Chapter 4
Congenital Lipoid Adrenal Hyperplasia

Walter L. Miller

Abstract All steroid hormones are made from cholesterol, which is primarily taken up by steroidogenic cells from circulating lipoproteins. The intracellular mechanisms by which cholesterol is delivered to cellular destinations remains under investigation. Steroidogenic cells must transport large amounts of cholesterol to, and then into, the mitochondria, where the cholesterol side-chain cleavage enzyme, P450scc, resides. P450scc is the first enzyme in steroidogenesis, converting insoluble cholesterol to soluble pregnenolone. The combination of genetic studies of a rare disease, congenital lipoid adrenal hyperplasia (lipoid CAH), and studies of the cell biology of mitochondrial cholesterol import led to the discovery of the steroidogenic acute regulatory protein (StAR). StAR acts exclusively on the outer mitochondrial membrane to trigger mitochondrial cholesterol import, but the precise mechanism of its action remains unclear. Lipoid CAH is caused by StAR mutations and is thus the StAR knockout experiment of nature. Recent work has shown that P450scc mutations cause a disease that is a phenocopy of lipoid CAH. The discovery of StAR led to the discovery of the broader family of structurally related StAR-related lipid transfer (START)-domain proteins. It appears that some of these START proteins may also be involved in intracellular cholesterol trafficking upstream from StAR.

Abbreviations

3βHSD	3β-hydroxysteroid dehydrogenase
ACAT	acyl-coenzyme A cholesteroacyl transferase
ACTH	adrenocorticotropic hormone
ANT	adenine nucleotide
CRAC	cholesterol recognition amino acid consensus domain
ER	endoplasmic reticulum
FAD	flavin adenine dinucleotide
HDL	high density lipoproteins
HSL	hormone-sensitive neutral lipase
HMGCoA	3-hydroxy-3-methylglutaryl co-enzyme A

W. L. Miller, M.D. (✉)
Department of Pediatrics, University of California San Francisco,
San Francisco, CA 94143-1346, USA
e-mail: wlmlab@ucsf.edu

B. J. Clark, D. M. Stocco (eds.), *Cholesterol Transporters of the START Domain Protein Family in Health and Disease,* DOI 10.1007/978-1-4939-1112-7_4,

IMM	inner mitochondrial membrane
IMS	intramembranous space
Km	Michaelis constant
LAL	lysosomal acid lipase
LDL	low-density lipoproteins
LH	luteinizing hormone
MENTAL	MLN64 N-terminal
MENTHO	MLN64 N-terminal domain homologue
MLN64	metastatic lymph node clone 64
NADPH	nicotinamide adenine dinucleotide phosphate
NPC	Niemann-Pick type C
OMM	outer mitochondrial membrane
PAP7	TSPO-associated protein 7 (ACBD3)
PBR	peripheral benzodiazepine receptor
PCP	phosphate carrier protein
PKA	protein kinase A
PKAR1A	protein kinase A regulatory subunit 1α
PRAX1	TSPO-associated protein 1
P450scc	mitochondrial cytochrome P450 specific for cholesterol side-chain cleavage
SF1	steroidogenic factor 1
SOAT	sterol O-acetyltransferase
SR-B1	scavenger receptor B1
StAR	steroidogenic acute regulatory protein
START	StAR-related lipid transfer domain
SREBPs	sterol regulatory element binding proteins
TSPO	18 kDa translocator protein,
VDAC1	voltage-dependent anion channel

Introduction: A Personal Historical Perspective on Lipoid CAH and StAR

Congenital lipoid adrenal hyperplasia (lipoid CAH) provides a perfect example of how study of a rare disease combined with contemporary biochemistry and genetics can open the door to an important new area of biology: the StAR-related lipid transfer domain (START) proteins. Early autopsy reports [1–4; Fig. 4.1] presaged the description of lipoid CAH as an inherited endocrine disorder, characterized by grossly enlarged, lipid-filled adrenals, disordered sexual development, and an apparent lack of adrenal steroidogenesis [5–7]. At that time, the enzymology of the conversion of cholesterol to pregnenolone was thought to require three enzymes, a 20α-hydroxylase, a 22-hydroxylase, and a 20, 22 "desmolase" [8]. Hence, the first case reported in English postulated that the disorder involved such an enzyme and

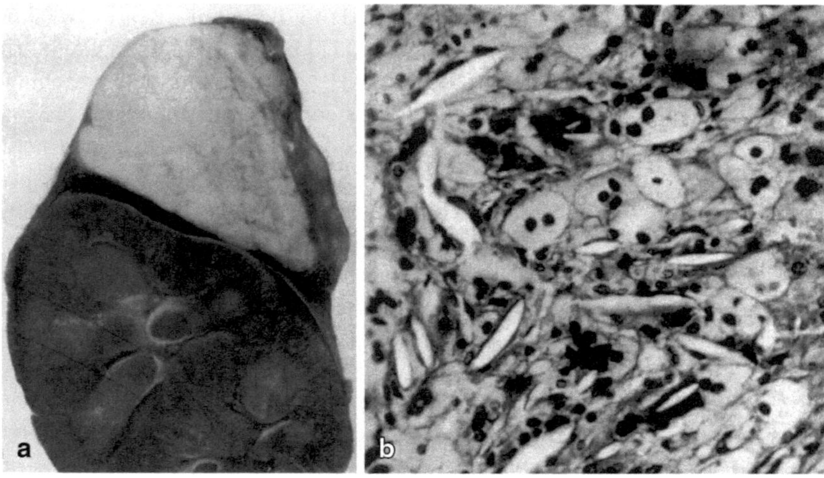

Fig. 4.1 Congenital lipoid adrenal hyperplasia (Lipoid CAH) from the 1955 report by Sandison. *Left*: Section through the autopsied kidney and adrenal of the 3 month female with probable lipoid CAH. *Right*: Histologic section stained with Hematoxolin and Eosin at 280× magnification, showing "small rather spindle-shaped cholesterol crystal clefts" (*Reproduced from Sandison AT. 1955. A form of lipoidosis of the adrenal cortex in an infant. Arch Dis Childh 30: 538–541, with permission from BMJ Publishing Group, Ltd.*)

named it "20, 22 desmolase deficiency" [9]. The first study of lipoid CAH in vitro found that mitochondria from affected tissue could convert 20α-hydroxycholesterol to pregnenolone, but not cholesterol to pregnenolone, suggesting that the defect was in a specific cholesterol 20α-hydroxylase [10]. Although this conclusion was incorrect, the experimental design was prescient: 23 years later, we used a similar approach to show that lipoid CAH was due to a mutation in the steroidogenic acute regulatory protein, StAR [11].

Advances in protein chemistry and the rapidly developing understanding of cytochrome P450 enzymes permitted the demonstration that cholesterol was converted to pregnenolone by a single mitochondrial cytochrome P450 enzyme, termed P450scc (where scc denotes cholesterol side chain cleavage; [12–14]. Others confirmed that lipoid CAH mitochondria could not convert cholesterol to pregnenolone but could not explain the efficacy of 20α-hydroxycholesterol as a substrate as reported by Degenhart [10]. Semi-quantitative CO-induced difference spectra indicated that lipoid CAH mitochondria had about half as much total P450 as control mitochondria, but had normal 11-hydroxylase activity, which was known to be catalyzed by a mitochondrial P450. Thus, this report concluded that lipoid CAH was caused by an absence of cytochrome P450scc [15], which remained the accepted explanation for this disease until 1991.

I first became aware of lipoid CAH in 1977 during my clinical fellowship in pediatric endocrinology, while taking care of a child with this disorder; several years later, I helped Bert Hauffa prepare a clinical report of this UCSF patient, reviewing the 34 cases then reported [16]. At that time I was developing a lab devoted to studying steroidogenesis, so I added lipoid CAH to the list of diseases we should try to "solve." Our cloning of human P450scc [17] permitted us to make the first, halting efforts at genetic studies of lipoid CAH with genomic deoxyribonucleic acid (DNA) from our UCSF patient [16]. We found no gene deletions [18], and with Ken Morohashi's cloning of the corresponding gene [19], we found normal P450scc exonic sequences. In collaboration with Paul Sanger, who was taking care of an affected family in New York, we obtained small amounts of affected testicular tissue from which we made ribonucleic acid (RNA), and found normal amounts of messenger RNA (mRNA) for P450scc and its electron-donating redox partners in affected gonadal tissue [20]. Thus, lipoid CAH was not due to a mutation in the enzyme system converting cholesterol to pregnenolone, suggesting a lesion in an upstream factor that delivered cholesterol to the mitochondria. The factors involved in this process were not understood at that time, and examination of then-contemporary candidates (sterol carrier protein 2, endozepine, GRP-78) did not yield an answer [20]. The work on endozapine by Hall and Shively [21] led us to the peripheral benzodiazepine receptor (PBR, now called mitochondrial transporter protein, TSPO), but in collaboration with Jerome F. Strauss 3rd (then at the University of Pennsylvania) we cloned this gene and found it was not mutated in lipoid CAH [22].

The factor that was missing in lipoid CAH had to have the three characteristics of the postulated "acute regulator" of steroidogenesis: (i) rapid inducibility by cyclic adenosine monophosphate (cAMP), (ii) short half-life evidenced by cycloheximide sensitivity, and (iii) capacity to induce steroidogenesis [23]. The work of Orme-Johnson [24–26] and Stocco [27] identified a family of peptides, termed pp30, pp32, and pp37 that appeared to have many of these properties. Barbara Clark in Doug Stocco's lab purified one of these, obtained a partial amino acid (AA) sequence, cloned it from mouse Leydig MA-10 cells, and found that its expression in MA-10 cells promoted steroidogenesis: the "steroidogenic acute regulatory protein," StAR, had been discovered [28].

Doug Stocco, Jerry Strauss, and my lab quickly joined forces. We had noted that placental steroidogenesis remained intact in lipoid CAH, indicating that the responsible gene had to be expressed in the adrenals and gonads, but not in the placenta [29]. I remember discussing this with Jerry on the phone in the summer of 1994 when he told me that Teruo Sugawara's northern blots showed that StAR was expressed in the adrenals and gonads but not in the placenta. We immediately concluded that StAR was the likely cause of lipoid CAH. Teruo and Jerry had cloned and were sequencing human StAR complementary DNA (cDNA) [30]. As soon as they had preliminary data from the 5′ and 3′ ends, we synthesized human StAR oligonucleotides and used polymerase chain reaction (PCR) to amplify the cDNA from the gonadal cDNA from Sanger's patient and from another patient from Alan Rogol in Virginia. Even before the full-length human cDNA sequence had been completed, we had found that StAR was mutated in lipoid CAH, thus establishing the basis of the disease. We then found mutations in the genomic DNA of these patients and our UCSF patient, thus firmly establishing the essential role of StAR by showing that lipoid CAH is a knockout experiment of nature [11]. Although it was clear that StAR mutations caused lipoid CAH, we needed an assay for StAR activity to show how StAR dysfunction caused the phenotype. We first did this by transfecting StAR cDNA expression vectors into COS-1 cells that had been transfected with vectors for human P450scc and its redox partners, adrenodoxin and adrenodoxin reductase, and measuring pregnenolone production [11]. However, because the activity of P450scc can vary with the abundance of these redox partners, we wanted to fix the ratio of these two proteins to eliminate assay variability. We did this with a catalytically active fusion protein of P450scc, adrenodoxin, and adrenodoxin reductase, which we had built previously [31, 32]. As a positive control we used 22-hydroxycholesterol, as by then it was known that soluble hydroxysterols are freely accessible to mitochondrial P450scc, bypassing the mitochondrial cholesterol import system [33]. Thus we were quickly able to establish the quantitative assay for StAR's action that remains in use today [11]. This experiment essentially recapitulated Degenhart's 1972 experiment: The 20-OH cholesterol they used bypassed the absent action of the then-unknown StAR protein, explaining why they found that this hydroxysterol, but not cholesterol, was converted to pregnenolone by lipoid CAH mitochondria. We soon found a splicing mutation in a fourth patient [34], and, as is so often the case, as soon as we had reported the cause of a rare disease, samples came in from around the world, permitting a general description of the genetics of lipoid CAH [35]. Thus, these early experiments proved that StAR enhanced cholesterol access to P450scc and that lipoid CAH served as a gene knockout of nature, proving its essential role. These early experiments also showed that there was a low level of StAR-independent steroidogenesis. This crucial observation led to our "two-hit model" that explains the clinical findings in lipoid CAH [35].

Delivery of Cholesterol to Mitochondria

The biosynthesis of all steroid hormones begins with the conversion of cholesterol to pregnenolone in mitochondria, followed by complex tissue-specific pathways leading to glucocorticoids and mineralocorticoids in the adrenal, estrogens and progestins in the placenta and ovary, and androgens in the testis [36]. Most steroidogenic tissues have specialized mechanisms to deliver cholesterol to mitochondrial P450scc. Lipoid CAH is one of several genetic disorders in these early steps in steroidogenesis, which have helped to elucidate this complex biology.

Intracellular cholesterol trafficking before cholesterol reaches steroidogenic mitochondria is essentially the same in steroidogenic and non-steroidogenic cells [37]. Cholesterol may be produced *de novo* from acetate via a complex pathway primarily found in the endoplasmic reticulum (ER) [38], but most steroidogenic cholesterol is imported from circulating lipoproteins. In rodents, the primary source is circulating high-density lipoproteins (HDL), which are imported into the cell by scavenger receptor B1 (SR-B1); in humans, the main source of steroidogenic cholesterol is low-density lipoproteins (LDL) (primarily derived from the diet), imported by receptor-mediated endocytosis via LDL receptors. However, patients with congenital abetalipoproteinemia, who have low LDL cholesterol, have normal basal cortisol concentrations and have only mildly impaired cortisol responses to adrenocorticotropic hormone (ACTH) [39]. Furthermore, treatment with high doses of statins does not impair cortisol secretion [40], hence endogenously produced cholesterol appears to be sufficient. LDL can suppress the rate-limiting enzyme in cholesterol synthesis, 3-hydroxy-3-methylglutaryl co-enzyme A (HMGCoA) reductase. A family of basic helix-loop-helix transcription factors called the sterol regulatory element binding proteins (SREBPs) regulates cholesterol uptake and its intracellular transport and utilization.

After an LDL particle is internalized by receptor-mediated endocytosis, the resulting endocytic vesicles fuse with lysosomes; the LDL proteins undergo proteolysis, and the liberated cholesteryl esters are hydrolyzed to "free" cholesterol by lysosomal acid lipase (LAL). However, the term "free cholesterol" is misleading as the solubility of cholesterol is only about 20 micromoles per liter; thus, "free cholesterol" refers to cholesterol that is bound to proteins or membranes, but lacks an ester covalently bound at carbon 3. Free cholesterol may be used by the cell or stored in lipid droplets following re-esterification by acyl-coenzyme A cholesterol-acyl-transferase (ACAT) also known as sterol O-acetyltransferase (SOAT1). HDL cholesteryl esters that enter the cell via SR-B1 are acted on by hormone-sensitive neutral lipase (HSL), following which the free cholesterol may similarly be used or re-esterified for storage. ACTH (in the adrenal) and luteinizing hormone (LH; in the gonad) increase intracellular cAMP, which stimulates HSL and inhibits ACAT, thus increasing the pool of free cholesterol available for steroidogenesis. Increasing cAMP stimulates HMGCoA reductase and LDL uptake. When intracellular cholesterol concentrations are high, transcription of the genes for LDL receptor, HMGCoA reductase, and LAL is repressed while ACAT is induced, thereby decreasing cholesterol uptake, synthesis, and de-esterification. Conversely, when intracellular cholesterol concentrations are low, this process is reversed [Fig. 4.2] [reviewed in 37].

Mutations in LAL cause Wolman disease and its milder variant, cholesterol ester storage disease [41]. Wolman disease is a lethal disorder characterized by visceral accumulation of cholesteryl esters and triglycerides, with secondary adrenal insufficiency. Affected infants develop hepatosplenomegaly, malabsorptive malnutrition, and developmental delay; adrenal insufficiency may be treated with hormonal-replacement therapy, but this has little impact on the relentless course of the disease, which is usually fatal in about a year. The diagnosis is suggested by calcifications that outline the adrenals (which is radiographically distinct from the gross adrenal hyperplasia of lipoid CAH) and is confirmed by finding deficient lysosomal acid lipase activity in leukocytes or fibroblasts. Bone marrow transplantation may ameliorate the disease by unclear mechanisms. In contrast to LAL, there is no known human HSL deficiency.

Endosomal/lysosomal cholesterol transport requires the NPC1 and NPC2 proteins that are mutated in Niemann-Pick type C (NPC) disease, characterized by endosomal accumulation of LDL-cholesterol and glycosphingolipids. Patients develop ataxia, dementia, speech loss, and spasticity at 2–4 years, and typically die at 10–20 years [42]. Cholesterol and other lipids accumulate in neurons, and there is robust glial infiltration. The diagnosis is made by finding characteristic foamy Niemann-Pick cells and "sea-blue" histiocytes in bone marrow. NPC2, a soluble

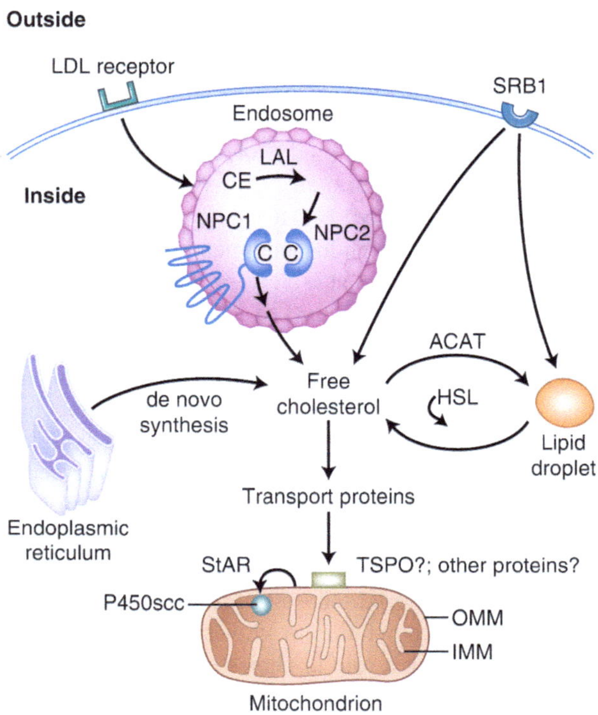

Fig. 4.2 Intracellular cholesterol trafficking. Human steroidogenic cells take up circulating low-density lipoproteins (LDL) by receptor-mediated endocytosis, directing the cholesterol to endosomes; rodent cells utilize cholesterol from high-density lipoproteins (HDL) via scavenger receptor B1 (SRB1). Cholesterol may also be synthesized from acetate in the endoplasmic reticulum (ER). Cholesteryl esters are cleaved by lysosomal acid lipase (LAL); free cholesterol is then bound by NPC2, transferred to NPC1, and exported. The metastatic lymph node clone 64/terminal domain homologue (MLN64/MENTHO) system resides in the same endosomes as the Niemann-Pick type C (NPC) system, but its role in cholesterol trafficking remains uncertain. Cholesterol may be re-esterified by acyl-CoA: cholesterol transferase (ACAT) and stored in lipid droplets as cholesteryl esters. Free cholesterol may be produced by hormone-sensitive lipase (HSL). Cholesterol can reach the outer mitochondrial membrane (OMM) by non-vesicular means by utilizing START-domain proteins or other cholesterol transport proteins. Movement of cholesterol from the OMM to the inner mitochondrial membrane (IMM) requires a multi-protein complex on the OMM. In the adrenals and gonads, the steroidogenic acute regulatory protein (StAR) is responsible for the rapid movement of cholesterol from the OMM to the IMM, where it can be converted to pregnenolone by P450scc (© *W.L. Miller*)

151 AA glycoprotein in the lysosomal lumen [43], binds cholesteryl esters with the cholesterol side chain oriented in a hydrophobic pocket and the polar 3βOH group exposed, allowing LAL to cleave the cholesteryl ester bond while bound to NPC2. Free cholesterol is then transferred to the N-terminal domain of NPC1, a 1278 AA glycoprotein with 13 transmembrane domains that span the endo-lysosomal membrane [44]. NPC1 binds cholesterol with the 3βOH group buried in the protein and

the side chain partially exposed; then inserts the cholesterol into the lysosomal membrane with the hydrophobic side chain going in first.

Two late endosomal proteins, metastatic lymph node clone 64 (MLN64) and MLN64 N-terminal domain homologue (MENTHO) may also participate in cholesterol trafficking [45]. MLN64 can bind cholesterol and co-localizes with NPC1 in late endosomes [46]. The N-terminal "MLN64 N-TerminAL" ("MENTAL") domain is structurally related to MENTHO [47], contains 4 transmembrane domains, and targets MLN64 to late endosomal membranes. The C-terminal domain of MLN64 is the START domain that is very similar to the lipid-binding domain of StAR [48, 49]. The MENTAL domains of MENTHO and MLN64 can interact to form homo- and heterodimers and to bind cholesterol, suggesting a role in endosomal cholesterol transport. MLN64 lacking the MENTAL domain (N-234 MLN64) has ~50–60% of StAR-like activity to stimulate mitochondrial cholesterol uptake [50, 51]. The START domain of MLN64 may interact with cytoplasmic HSP60 to stimulate steroidogenesis in placental mitochondria [52]. An essential function for MLN64 is not established as knockout of the START domain of MLN64 yields viable, neurologically intact, fertile mice with normal plasma and hepatic lipids [53]. Human genetic disorders of MLN64 or MENTHO are not known. Accumulation of cholesterol in NPC1-deficient cells increases MLN64-mediated cholesterol transport to mitochondria and accumulation of cholesterol in the outer mitochondiral membrane (OMM), suggesting a role in cholesterol transport from endosomes to mitochondria [54].

Intracellular cholesterol transport may be "vesicular" (mediated by membrane fusion) or "non-vesicular" (bound to proteins). Both vesicular and non-vesicular cholesterol transport occur in steroidogenic cells, but non-vesicular transport involving high-affinity cholesterol-binding START-domain proteins appears to be the principal means for cholesterol transport to mitochondria [37]. START domain proteins are found in all eukaryotes; the 15 mammalian START proteins are termed STARD1–15 (StAR itself is also known as STARD1, but is designated in this chapter by its classical name, StAR) [49]. STARD4, D5, and D6, the START proteins most closely related to StAR, bind cholesterol, are induced by SREBP, and lack N-terminal signal sequences, suggesting they are cytosolic sterol transport proteins [37]. StAR, STARD3 (MLN64), D4, and D5 bind cholesterol with high-affinity and specificity, facilitate cholesterol transport, and appear to play roles in cellular cholesterol homeostasis [55]. STARD5 is predominantly expressed in Kupffer cells, macrophages, and proximal renal tubules and appears to act primarily as a bile-acid-binding protein [56, 57]. StAR and STARD6 stimulate the movement of cholesterol from the OMM to the inner mitochondrial membrane (IMM), but STARD4 and STARD5 do not [58]. STARD4 and/or D5 may bring cholesterol to the OMM; however, STARD4 knockout mice have no changes in steroidogenesis and minimal changes in weight and serum lipids, hence an essential function of STARD4 is not apparent [59]. Thus, it appears that the principal mechanism for getting cholesterol to the mitochondria is by non-vesicular transport involving proteins that are structurally related to StAR, but that StAR itself plays a minor role in this step. The OMM of adrenal mitochondria contains abundant cholesterol, whereas the IMM contains relatively little cholesterol. Whether all OMM cholesterol is available for steroidogenesis is unclear. Early studies identified a distinct pool of "steroidogenic" OMM cholesterol that was distinct from the structural membrane cholesterol and could be mobilized by cAMP [60]. Whether there are two kinetically distinct pools of cholesterol in the OMM, and how cholesterol is transferred to the IMM remain under investigation.

Entry of Cholesterol into Steroidogenic Mitochondria: Action of the Steroidogenic Acute Regulatory Protein, StAR

Unlike cells that produce polypeptide hormones, which store mature hormone in secretory vessicles for rapid release, steroidogenic cells store very little steroid, so that steroid secretion requires more steroid synthesis, which can be induced by several mechanisms [36, 37]. In the adrenal, ACTH promotes steroidogenesis at three distinct levels. First, over the course of months, ACTH stimulates adrenal growth via several growth factors; second, over the course of days, ACTH, in the adrenal zona fasciculata and angiotensin II in the zona glomerulosa stimulate transcription of steroidogenic enzyme genes, especially the *CYP11A1* gene encoding P450scc, thus increasing the amount of steroidogenic machinery; third, within an hour, ACTH stimulates the activation of preexisting StAR (by its phosphorylation at Ser195) and the synthesis of new StAR. StAR then appears to interact with a macromolecular complex on the OMM to increase cholesterol flux from the OMM to the IMM, where it becomes the substrate for P450scc. The first two modes of steroid regulation comprise the chronic steroidogenic response and the action of StAR comprises the acute steroidogenic response [36, 37].

Co-expression of StAR and P450scc in nonsteroidogenic cells increases conversion of cholesterol to pregnenolone, suggesting that StAR triggers the acute steroidogenic response [23]. The indispensible role of StAR was established by finding that StAR mutations cause lipoid CAH [11, 34, 35]. However, some steroidogenesis can take place without StAR: The human placenta synthesizes steroids via P450scc but expresses little or no StAR [30], and cells expressing P450scc but not StAR can convert cholesterol to pregnenolone [31, 32] at ~14 % of the maximal StAR-induced rate [11, 35]. The mechanisms underlying this StAR-independent steroidogenesis remain unclear. The placenta produces a cleavage product of MLN64 that appears to have StAR-like activity [51], or placental StAR-independent activity might happen without a transport protein, especially if soluble oxysterols are the substrate. The crystal structure of the START domain of MLN64 [61], computational models of StAR [62, 63], and low-resolution crystallography of StAR [64] all show a globular protein with an α/β helix-grip fold and an elongated hydrophobic pocket that accommodates one cholesterol molecule with its 3β-OH group coordinated by the two polar residues. These structures and the crystal structure of STARD4 [65] are characterized by long α-helixes at the N- and C-termini, two short α-helixes, and nine antiparallel β sheets that form a helix-grip fold.

How StAR induces the import of cholesterol into mitochondria remains unclear. StAR is synthesized as a 37-kDa protein with an N-terminal mitochondrial leader sequence that is cleaved during mitochondrial import to yield 30-kDa intramitochondrial StAR. The 37 kDa cytoplasmic form is often termed a "precursor" and the 30 kDa intramitochondrial protein its "mature form," but these terms do not describe the biology of StAR [66]. Moreover, deletion of the leader peptide has no effect on its activity [67]. Mitochondrial localization experiments show that StAR is active on the OMM but not in the IMS or on the matrix side of the IMM, and experiments manipulating the speed of StAR's mitochondrial entry show that faster import decreases activity and slower import increased activity [68]. Thus, StAR acts exclusively on the OMM and its activity is proportional to how long it remains on the OMM, so that it is the OMM localization of StAR, and not its cleavage from the 37 kDa form to the 30 kDa form, that determines its activity.

Mechanism of StAR's Action—An Ongoing Study

The mechanism by which StAR triggers cholesterol flux from the OMM to the IMM requires pH-induced conformational changes in StAR that are needed for StAR to accept and discharge cholesterol [69, 70]. Cholesterol is blocked from reaching StAR's binding pocket by a set of hydrogen bonds that immobilize the C-helix but are disrupted when the surface residues of StAR are protonated as happens when StAR interacts with charged phospholipids on the OMM, thus eliciting a conformational change that permits cholesterol access [70]. Thus, the activity of StAR on the OMM requires an acid-induced disruption of hydrogen bonds and a consequent conformational change in StAR to permit it to bind and release cholesterol. StAR can transfer cholesterol between synthetic, protein-free membranes in vitro, but with non-physiologic stoichiometry [71], and the biologically inactive StAR mutant R182L can transfer cholesterol to membranes in vitro [72], indicating that cholesterol binding is necessary but not sufficient for StAR activity. The phosphorylation of StAR on Ser 195 doubles its activity [73]. It appears that the protein kinase A (PKA) anchor protein AKAP121 recruits the type II PKA regulatory subunit α (PKAR2A) to the OMM, which phosphorylates StAR, whereas the type I kinase drives StAR transcription [74]. Interaction between the 37 kDa StAR and HSL has also been reported [75]. StAR is recycled as each molecule moves hundreds of molecules of cholesterol into the mitochondria before it is inactivated by mitochondrial import [68, 76]. Low levels of steroidogenesis will persist in the absence of StAR at about 14% of the StAR-induced rate [11, 35], accounting for the steroidogenic capacity of tissues that lack StAR, such as the human placenta.

StAR appears to interact with a multi-protein complex on the OMM. In addition to AKAP121 and PKAR2A mentioned above, several other proteins have been implicated, but what role each plays in the mitochondrial importation of cholesterol remains unclear [77–79]. The first identified and most extensively studied of these proteins is the PBR, now also called TSPO [80]. PBR/TSPO is a ubiquitously expressed 18 kDa mitochondrial protein; based on the binding of drug ligands, it is estimated that it comprises about 2% of adrenal OMM protein [81, 82]. PBR/TSPO lacks a mitochondrial targeting sequence, but its C-terminal half targets it to the OMM [83]. Ligands of PBR/TSPO appear to stimulate cholesterol movement from OMM to IMM and stimulate steroidogenesis [84]. PBR/TSPO is a component of a 140–200 kDa complex consisting of PBR/TSPO, the 34-kDa voltage-dependent anion channel (VDAC1), the 30-kDa the adenine nucleotide transporter (ANT), the 10-kDa diazepam-binding inhibitor (acyl-CoA-binding domain 1, ACBD1), the TSPO-associated protein-1 (PRAX-1), and the PKA regulatory subunit RIα-associated protein 7 (PAP7) [78]. PAP7, also known as acyl-CoA binding domain-containing protein 3 (ACBD3), appears to bind both TSPO and regulatory subunit RIα (and RII to a lesser extent) of cAMP-dependent protein kinase A (PKA) [85]. PBR/TSPO interacts with VDAC on the OMM, possibly anchoring the multi-protein complex to the OMM and assisting the binding and import of StAR [86]; VDAC1 also interacts with ANT on the IMM [87]. While StAR and PBR/TSPO appear to interact functionally [88], no physical interaction has been found; protein cross-linking experiments identify contacts between StAR and VDAC1 and phosphate carrier protein (PCP), but not PBR/TSPO [77]. VDAC1 interacts with PBR/TSPO [83, 86], but how VDAC-1 might participate in cholesterol import is unclear as VDAC1 has a cylindrical structure with a hydrophilic interior for anion transport but is ill-suited as a channel for hydrophobic cholesterol [88, 89]. VDAC is found at contact sites between the OMM and the IMM [90] where it may complex with hexokinase, ANT, creatine kinase, and proteins of the Bcl-2 family [91]. PBR/TSPO has five transmembrane α-helices spanning the OMM [92], suggesting it functions as a cholesterol channel that would act downstream from StAR. PBR/TSPO has a cytoplasmic domain containing a "cholesterol recognition amino acid consensus" (CRAC) domain that binds cholesterol, suggesting that this domain participates in transferring cholesterol from the OMM to the IMM [93]. Mutagenesis of the CRAC domain interferes with cholesterol binding and cholesterol transfer to the IMM, and blocking the binding of cholesterol to CRAC prevents steroidogenesis [94]. PBR/TSPO knockdown in Leydig cells disrupts cholesterol transport and steroidogenesis [95]. It has been reported that disruption of the TSPO gene in Leydig cells blocked cholesterol transport into the mitochondria and reduced steroid production [96], and that PBR/TSPO knockout mice experience embryonic lethality [84], suggesting an indispensible role in development. However, in a recent report, tissue-specific knockout of PBR/TSPO in mouse Leydig cells yielded no reproductive or steroidogenic phenotype [97]. Thus, PBR/TSPO joins MLN64 [53] and StARD4 [59] on a growing list of proteins for which there is good biochemical evidence for a role in steroidogenesis, but for which the mouse knockout data do not confirm a role, and for which no definitive answer is known from a human genetic disease. Thus, the role of PBR/TSPO in steroidogenesis remains controversial.

Mutations in StAR—Congenital Lipoid Adrenal Hyperplasia (Lipoid CAH)

As reviewed in the Introduction, and in greater detail elsewhere [98], lipoid CAH is a rare disorder that has played a central role in driving contemporary understanding of the regulation of steroidogenesis and the discovery of the START-domain proteins. Lipoid CAH is the most severe genetic disorder of steroid hormone synthesis, characterized by absent or very low serum concentrations of all steroids, high basal ACTH and plasma renin activity, an absent steroidal response to long-term treatment with high doses of ACTH or hCG, and grossly enlarged adrenal glands filled with cholesterol and cholesteryl esters. Although these findings initially suggested a lesion in the conversion of cholesterol to pregnenolone, the *CYP11A1* gene for P450scc is not mutated in these patients, and the mRNAs for adrenodoxin reductase and adrenodoxin are intact [20]. Furthermore, the placenta (a fetal tissue) continues to produce progesterone in lipoid CAH, proving that the cholesterol side chain cleavage system remains intact, and permitting normal term gestation [29]. The normal P450scc system plus the accumulation of cholesterol esters in the affected adrenal suggested that the lesion lay in a factor upstream from P450scc, such as in a factor involved in mitochondrial cholesterol transport [20]; mutations in PBR/TSPO were sought and excluded [22]. Thus, the responsible factor appeared to be expressed in the adrenal and gonad, but not in the placenta, and the discovery that StAR expression had this tissue distribution helped to identify it as the disordered step in lipoid CAH [11].

Lipoid CAH is the StAR gene knockout experiment of nature, revealing the complex physiology of the StAR protein [98]. StAR promotes steroidogenesis by increasing the movement of cholesterol into mitochondria, but in the absence of StAR, steroidogenic cells still make small amounts of steroids by StAR-independent steroidogenesis [35, 37, 68]. This observation led to the two-hit model of lipoid CAH [35] [Fig. 4.3]. The first hit is the loss of StAR itself, leading to a loss of most, but not all steroidogenesis, leading to a compensatory rise in ACTH and LH. These increased tropic hormones increase production of intracellular cAMP, which increases biosynthesis of LDL receptors, their consequent uptake of LDL cholesterol, and *de novo* synthesis of cholesterol. In the absence of StAR, this increased intracellular cholesterol accumulates as in a storage disease, causing the second hit, which is the mitochondrial and cellular damage caused by the accumulated cholesterol, cholesterol esters, and their autooxidation products [35, 37].

The two-hit model explains the unusual clinical findings in lipoid CAH. In response to tropic stimulation by hCG, fetal Leydig cells normally make large amounts of testosterone; in lipoid CAH, this stimulation leads to accumulation of cholesterol and cellular destruction early in gestation. This eliminates the testosterone biosynthesis needed for development of male external genitalia, so that an affected 46, XY fetus does not undergo normal virilization and is born with female-appearing external genitalia and a blind vaginal pouch. However, Wolffian duct derivatives (seminal vesicle, ejaculatory ducts) are well developed [99], indicating the presence of some testosterone synthesis in early fetal life as predicted by the two-hit

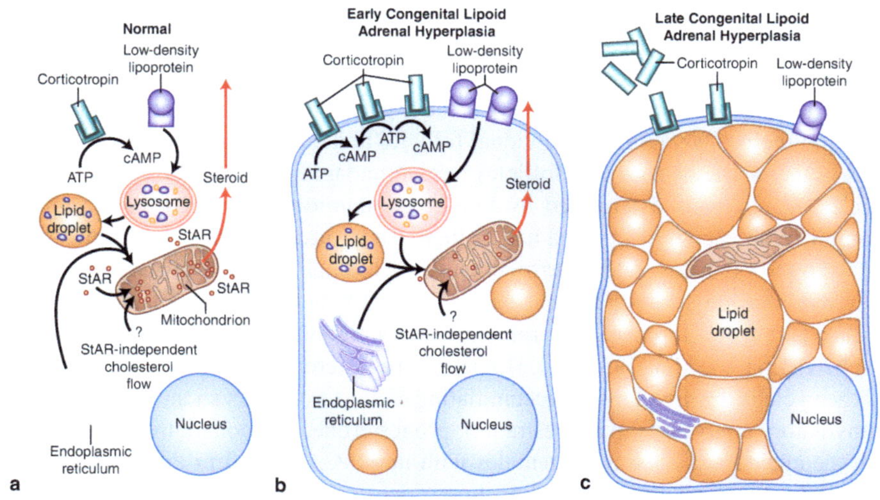

Fig. 4.3 Two-hit model of congenital lipoid adrenal hyperplasia (lipoid CAH). **a** In normal adrenal cells, cholesterol is primarily derived from low-density lipoproteins, and the rate-limiting step in steroidogenesis is movement of cholesterol from the outer mitochondrial membrane (OMM) to the inner mitochondrial membrane (IMM). **b** Early in lipoid CAH, StAR independent steroidogenesis moves small amounts of cholesterol into mitochondria, yielding sub-normal steroidogenesis; adrenocorticotropic hormone (ACTH) secretion increases, stimulating further accumulation of cholesteryl esters in lipid droplets. **c** As lipids accumulate, they damage the cell through physical engorgement and by the action of cholesterol auto-oxidation products; steroidogenic capacity is destroyed, but tropic stimulation continues. Ovarian follicular cells remain unstimulated and undamaged until puberty, when small amounts of estradiol are produced, as in **b**, causing phenotypic feminization, with infertility and hypergonadotropic hypogonadism. Modified from [35]

model. The undamaged Sertoli cells function normally and produce Müllerian inhibitory hormone, so that the phenotypically female 46, XY fetus with lipoid CAH has no cervix, uterus, or fallopian tubes. The steroidogenically active fetal zone of the adrenal is similarly affected, eliminating most dehydroepiandrosterone (DHEA) biosynthesis, thus eliminating the feto-placental production of estriol, so that mid-gestation maternal and fetal estriol levels are very low [29]. The definitive zone of the fetal adrenal, which differentiates into the zonae glomerulosa and fasciculata, normally produces very little aldosterone. Because fetal salt and water homeostasis is maintained by the placenta, stimulation of the glomerulosa by angiotensin II generally does not begin until birth, so that the glomerulosa does not necessarily experience the "second hit" during fetal life. Consequently, many newborns with lipoid CAH do not have a salt-wasting crisis until after several weeks of life when chronic stimulation then leads to cellular damage [35, 100].

The two-hit model also explains the spontaneous feminization of affected 46, XX females who receive glucocorticoid and mineralocorticoid replacement therapy in infancy and hence reach adolescence [35, 101, 102]. The fetal ovary makes little or no steroids and contains no steroidogenic enzymes after the first trimester [103]; consequently, the ovary remains largely undamaged until it is stimulated by gonadotropins at the time of puberty when it then produces small amounts of estrogen by StAR-independent steroidogenesis. Continued stimulation results in cholesterol accumulation and cellular damage, so that there is minimal biosynthesis of progesterone in response to the LH surge in the latter part of the cycle. Because gonadotropin stimulation only recruits individual follicles and does not promote steroidogenesis in the whole ovary, most follicles remain undamaged and available for future cycles. Cyclicity is determined by the hypothalamic-pituitary axis, and remains normal. With each new cycle, a new follicle is recruited and more estradiol is produced by StAR-independent steroidogenesis. Although net ovarian steroidogenesis is impaired, enough estrogen is produced (especially in the absence of androgens) to induce breast development, general feminization, monthly estrogen withdrawal and cyclic vaginal bleeding [35, 101]. However, progesterone synthesis in the latter half of the cycle is disturbed by the accumulating cholesterol esters so that the cycles are anovulatory. Measurements of estradiol, progesterone, and gonadotropins throughout the cycle in affected adult females with lipoid CAH confirm this model [102]. Similarly, examination of StAR-knockout mice confirms the two-hit model [104]. Thus, examination of patients with lipoid CAH has elucidated the physiology of the StAR protein in each steroidogenic tissue.

Genetic analysis of patients with lipoid CAH has revealed numerous mutations in the StAR gene [35, 37, 66]. Lipoid CAH is relatively common in Japan and Korea, where the carrier frequency is approximately one in 300, so that one in every 250,000 to 300,000 newborns in these countries is affected for a total of about 500 patients in Japan and Korea. About 65–70 % of affected Japanese alleles and about 90 % of affected Korean alleles carry the mutation Q258X [35, 105, 106], which was identified in the first patients studied [11]. Other genetic clusters are found among Palestinian Arabs, most of whom carry the mutation R182L [35], in eastern Saudi Arabia, carrying R188C [100], and in Switzerland, carrying the mutation L260P [107]. Deletion of only 10 carboxy-terminal residues reduces StAR activity by half, and deletion of 28 carboxy-terminal residues by the common Q258X mutation eliminates all activity. By contrast, deletion of the first 62 amino-terminal residues has no effect on StAR activity, even though this deletes the entire mitochondrial leader sequence and forces StAR to remain in the cytoplasm [67]. Physical studies and partial proteolysis indicate that residues 63–193 of StAR (i.e., the domain that lacks most of the crucial residues identified by missense mutations) are protease-resistant and constitute a "pause-transfer" sequence, which permits the bioactive loosely folded carboxy-terminal molten globule domain to have increased interaction with the outer mitochondrial membrane [69].

The clinical findings in most patients with lipoid CAH are quite similar: An infant with normal-appearing female genitalia experiences failure to thrive and salt loss in the first weeks or months of life and is diagnosed with adrenal insufficiency because of hyponatremia and hyperkalemia and possibly also because of hyperpigmentation [35–37]. However, other clinical presentations are possible, including apparent sudden infant death syndrome (SIDS) [108] and late initial presentation of salt loss at about one year of age [100]. We also described an attenuated form called "non-classic lipoid CAH," which is caused by mutations that retain about 20–25 % of normal StAR activity [109]. These are usually children who first experience symptoms of adrenal insufficiency after several years, and the 46, XY patients have normal-appearing male external genitalia. These patients typically have minimal impairments in mineralocorticoid synthesis, evidenced by elevated plasma renin activity with normal serum sodium and potassium, and may also have mild hypergonadotropic hypogonadism. Some patients have been diagnosed in adulthood and have been mistaken for having "familial glucocorticoid deficiency," a blanket term referring to disorders of ACTH action, but elevated plasma renin, hypergonadotropic hypogonadism, and azospermia are not features of familial glucocorticoid deficiency. Most of these patients carry the StAR mutation R188C [110, 111]. Thus, the spectrum of clinical presentations of StAR defects is substantially broader than was initially appreciated from studies of classic lipoid CAH.

Treatment of lipoid CAH consists of hormonal replacement therapy, and can be thought of in three phases. First, at time of diagnosis, most infants are hypovolemic, hyponatremic, hyperkalemic, and acidotic as a consequence of both mineralocorticoid and glucocorticoid deficiency. Initial treatment typically consists of fluid recussitation, salt replacementation to replace lost sodium, mineralocorticoid therapy (fludrocortisol) to replace lost mineralocorticoid synthesis, and initial stress-dosing with glucocorticoids to deal with physiologic stress and to suppress the hypothalamic-pituitary-adrenal axis [16, 112]. Second, once the life-threatening peri-diagnostic period has been dealt with and the intravascular volume and salt deficit have been repaired, physiologic replacement doses of glucocorticoids and mineralocorticoids will permit normal growth and development to adulthood. The glucocorticoid requirement in lipoid CAH is analagous to that in Addison disease and is less than in the virilizing forms of congenital adrenal hyperplasias due to 21-hydroxylase or 11-hydroxylase deficiency because adrenal over-suppression of excess adrenal androgen production is not needed. With glucocorticoid supplementation confined to physiologic oral replacement doses of 6–8 mg m^2/d [113–115], growth in these patients should be normal, although ACTH may remain elevated even when supraphysiologic doses of glucocorticoids are given [16]. Severely affected 46, XY newborns have normal female external genitalia and should undergo orchiectomy later in life and be raised as females, with estrogen replacement therapy started at the usual age of puberty to permit feminization of the body habitus and prevent bone loss. Third, affected 46, XX females, typically have spontaneous pubertal feminization but anovulatory cycles and early secondary amenorrhea, also requiring sex hormone replacement therapy. Successful pregnancy has been achieved in an adult female with lipoid CAH by clomiphene citrate stimulation followed by progesterone supplementation to mimic the maternally produced first trimester progesterone that the affected mother could not produce [116].

P450scc: Cholesterol's Mitochondrial Destination

Mutations in other genes can produce a clinical phenotype that is essentially the same as that caused by StAR mutations. Because these disorders derive from different molecular lesions, they should not be called lipoid CAH. Although the initial

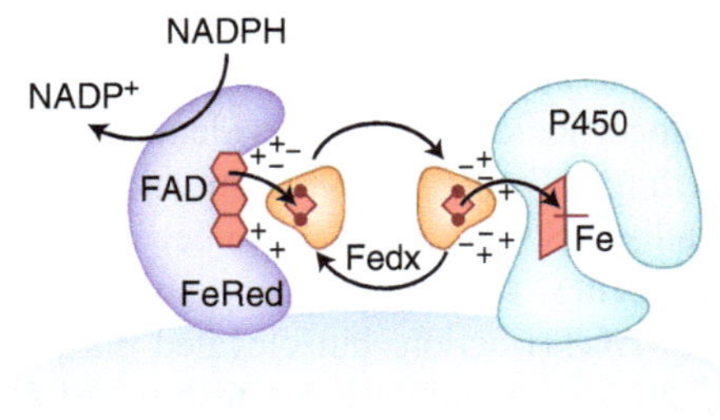

Fig. 4.4 Organization of mitochondrial P450 enzyme systems. Nicotinamide adenine dinucleotide phosphate (NADPH) first donates electrons to the flavin adenine dinucleotide (FAD) moiety of ferredoxin reductase (FeRed); ferredoxin reductase then interacts with ferredoxin (Fedx) by charge-charge attraction, permitting electron transfer of the Fedx to the Fe_2S_2center (ball and stick diagram). Ferredoxin then dissociates from ferredoxin reductase and diffuses through the mitochrondrial matrix. The same surface of ferredoxin that received the electrons from ferredoxin reductase then interacts with the redox-partner binding-site of a mitochondrial P450, such as P450scc, and the electrons then travel to the heme ring of the P450. The heme iron then mediates catalysis with substrate bound to the P450. © W.L. Miller

genetic characterizations of patients with apparent lipoid CAH found no patients with mutations in the *CYP11A1* gene encoding P450scc, the first such patient was described in 2001 [117]. The physiology of these patients requires knowledge of the biochemistry of P450scc.

Once cholesterol reaches the IMM, it may be converted to pregnenolone by the cholesterol side chain cleavage enzyme, P450scc, to initiate steroidogenesis. Most steroidogenic enzymes are cytochrome P450 enzymes, all of which have approximately 500 residues, contain a single heme group, and absorb light at 450 nm when reduced with carbon monoxide. The human genome contains 57 *CYP* genes encoding cytochrome P450 enzymes. Seven human cytochrome P450 enzymes are targeted to the mitochondria, with the other 50 being targeted to the ER; the roles of the human P450 enzymes have been reviewed recently [118]. P450 enzymes use their heme iron to activate molecular oxygen using electrons donated by nicotinamide adenine dinucleotide phosphate (NADPH). Mitochondrial P450 enzymes such as P450scc are designated as Type 1 P450 enzymes, and receive electrons from NADPH via an electron transfer chain consisting of a flavoprotein termed ferredoxin reductase and a small iron-sulfur protein termed ferredoxin [119] [Fig. 4.4]. The type 2 P450 enzymes in the ER receive electrons via the single 2-flavin protein P450 oxidoreductase [119]. All P450 enzymes can catalyze multiple chemical reactions, often with very different substrates.

Conversion of cholesterol to pregnenolone by P450scc is the first, rate-limiting and hormonally regulated step in the synthesis of all steroid hormones. P450scc catalyzes the 22-hydroxylation of cholesterol, 20-hydroxylation of 22(R)-hydroxycholesterol, and oxidative scission of the C20–22 bond of 20(R), 22(R)-dihydroxycholesterol (the side chain cleavage event), yielding pregnenolone and isocaproaldehyde. The binding of cholesterol and the initial 22-hydroxylation are rate-limiting as the efficiencies (k_{cat}/Km ratios) are much higher for the subsequent reactions, and the high K_D of ~3000 nM drives the dissociation of pregnenolone from P450scc [36]. Alternatively, soluble hydroxysterol substrates such as 22(R)-hydroxycholesterol can enter the mitochondrion without the action of StAR. The conversion of cholesterol to pregnenolone is slow with a net turnover number of about 6 to 20 molecules of cholesterol per molecule of P450scc per second [14, 120]. Because 20-hydroxycholesterol, 22-hydroxycholesterol, and 20, 22-hydroxycholesterol can all be isolated from bovine adrenals, and because 3 moles of NADPH are required per mole of cholesterol converted to pregnenolone, it was initially thought that three separate enzymes were involved. However, protein purification and reconstitution of enzymatic activity in vitro showed that a single protein, P450scc, converts cholesterol to pregnenolone [reviewed by 37]. The crystal structures of bovine [121] and human [122] P450scc, the latter in complex with ferredoxin show that the single active site of P450scc is in contact with the IMM. It is the expression of the *CYP11A*1 gene that renders a cell "steroidogenic." Pregnenolone appears to exit the mitochondrion unaided; no transport protein has been found, and physiologic evidence does not suggest the presence of such a transporter.

Catalysis by P450scc requires two electron-transfer intermediates, ferredoxin reductase and ferredoxin [reviewed in 119]. Ferredoxin reductase receives electrons from NADPH then forms a 1:1 complex with ferredoxin which then dissociates and forms an analogous 1:1 complex with P450scc or other mitochondrial P450 enzymes, thus functioning as an indiscriminate, diffusible electron shuttle for all mitochondrial P450s (Fig. 4.4). In addition to the inherent properties of P450scc, the abundances of ferredoxin reductase and ferredoxin determine catalytic activity [31]. Genetic disorders of human ferredoxin reductase and ferredoxin have not been described, and mouse knockouts have not been reported. Mutation of the Drosophila ferredoxin reductase homologue *dare* causes developmental arrest and degeneration of the adult nervous system secondary to disrupted ecdysone production [123]. In vitro, the requirement for ferredoxin reductase and ferredoxin is not absolute as P450scc fused to microsomal P450 oxidoreductase remains active when targeted to the mitochondria, but is inactive when targeted to the ER even when supplied with the 22(R)-hydroxycholesterol substrate that bypasses the StAR system [32]. Thus, the mitochondrial localization is essential for the enzymatic activity of P450scc.

P450scc Deficiency

The suppression of uterine contractility by progesterone prevents spontaneous abortion and is essential for the maintenance of mammalian pregnancy. However, different species use different strategies for the production of this progesterone: In rodents and ungulates, the source is maternal as the maternal corpus luteum of pregnancy continues to produce progesterone to term, whereas in primates, the source is a fetal tissue, the placenta. Thus, the available models of defective P450scc function, a spontaneously occurring *CYP11A1* deletion in the rabbit [124] and knockout of the *CYP11A1* gene in the mouse [125] are not informative about the potential consequences of disorders in human P450scc. Once StAR mutations were found to cause lipoid CAH [11, 34, 35], it seemed logical that the dependence of human pregnancy on placental progesterone would be incompatible with P450scc-deficient fetuses surviving to term [126]. Nevertheless, beginning in 2001 [117], patients were reported who had clinical and hormonal findings that were indistinguishable from those of patients with StAR mutations, but who had mutations in P450scc. To date, 19 such patients have been described, but no patient with a P450scc mutation has had the adrenal hyperplasia typically seen in lipoid CAH [127]. Most of these patients had mutations that ablated all P450scc activity. It is not clear how these fetuses with P450scc mutations reached term gestation; one possibility is that these pregnancies were carried to term because of unusually protracted maintenance of the maternal corpus luteum of pregnancy, which normally involutes in the second trimester, but this has not been investigated directly. These patients may be clinically indistinguishable from those with lipoid CAH, and are treated in exactly the same fashion. The 46, XY genetic males fail to produce testosterone during fetal life, and are born with female external genitalia although their internal reproductive structures are male as their testes produced anti-Müllerian hormone. Following birth, these patients require steroid hormone replacement therapy and may have long-term survival. As with non-classical lipoid CAH, a milder "non-classical" form of P450scc deficiency that is clinically and hormonally indistinguishable from non-classic lipoid CAH has been reported in patients with P450scc mutations that retain 10–20% of wild-type activity [128, 129]. No hormonal test distinguishes lipoid CAH from P450scc deficiency, but the grossly enlarged adrenals that give lipoid CAH its name have not been described in patients with P450scc mutations, sometimes permitting radiologic distinction, but the only definitive test to distinguish these disorders is DNA sequencing [130].

Another disorder occasionally mistaken for lipoid CAH is deficiency of steroidogenic factor 1 (SF1), a transcription factor required for adrenal and gonadal, but not for placental, expression of genes for the steroidogenic enzymes [131]. More than 50 patients have been described carrying SF1 mutations [132, 133]. The phenotype of SF1-deficient patients is variable; some are 46, XY with a female phenotype and adrenal failure, thus resembling lipoid CAH, but in most cases, the gonadal phenotype predominates and there is little if any impairment of adrenal steroidogenesis. The Leydig cells may have lipid accumulation and progressive degeneration similar to the findings in lipoid CAH [133].

References

1. Tilp A. Hochgradige Verfettung der Nebenniere eines Sauglings. Verhandlungen der Deutschensch Gesellschaft fur Pathologie.1913;16:305–7.
2. Brutschy P. Hochgradige Lipoidhyperplasie beider Nebennieren mit herdformigen Kalkablagerunger bei einem Fall von Hypospadiasis penisscrotalis und doppelseitigem Kryptorchismus mit unechter akzessorischer Nebenniere am rechten Hoden (Pseudohermaphroditismus masculinus externus). Frankfurter Zeitschriff für Pathologie. 1920;24:203–40.

3. Zahn J. Ueber intersexualitat und Nebennierenhyperplasie. Schweiz Med Wochenschr. 1948;78:480–6.
4. Sandison AT. A form of lipoidosis of the adrenal cortex in an infant. Arch Dis Childh. 1955;30:538–41.
5. Prader A, Gurtner HP. Das Syndrom des Pseudohermaphroditismus masculinus bei kongenitaler Nebennierenrindenhyperplasie ohne Androgenuberproduktion (adrenaler Pseudohermaphroditismus masculinus). Helvetica Paediatrica Acta. 1955;10:397–412.
6. Prader A, Siebenmann RE. Nebenniereninsuffzienz bie kongenitaler Lipoidhyperplasie der Nebennieren. Helvetica Paediatrica Acta. 1957;12:569–95.
7. Prader A, Anders CJPA. Zur Genetik der kongenitalen Lipoidhyperplasie der Nebennieren. Helvetica Paediatrica Acta. 1962;17:285–9.
8. Shimizu K, Hayano M, Gut M, Dorfman RI. The transformation of 20α; hydroxycholesterol to isocaproic acid and C21 steroids. J Biol Chem. 1961;236:695–9.
9. Camacho AM, Kowarski A, Migeon CJ, Brough A. Congenital adrenal hyperplasia due to a deficiency of one of the enzymes involved in the biosynthesis of pregnenolone. J Clin Endocrinol Metab. 1968;28:153–61.
10. Degenhart HJ, Visser KHA, Boon H, O'Doherty NJD. Evidence for deficiency of 20α cholesterol hydroxylase activity in adrenal tissue of a patient with lipoid adrenal hyperplasia. Acta Endocrinol. 1972;71:512–8.
11. Lin D, Sugawara T, Strauss JF 3rd, Clark BJ, Stocco DM, Saenger P, Rogol A, Miller WL. Role of steroidogenic acute regulatory protein in adrenal and gonadal steroidogenesis. Science. 1995;267:1828–31.
12. Simpson ER, Boyd GS. The cholesterol side-chain cleavage system of bovine adrenal cortex. Euro J Biochem. 1967;2:275–85.
13. Shikita M, Hall PF. Cytochrome P-450 from bovine adrenocortical mitochondria: An enzyme for the side chain cleavage of cholesterol. I. Purification and properties. J Biol Chem. 1973;248:5596–604.
14. Katagiri M, Takemori S, Itageki E, Suhara K, Gomi T, Sato H. Characterization of purified cytochrome P450scc and P450-11β from bovine adrenocortical mitochondria. Adv Exp Med Biol. 1976;74:281−95.
15. Koizumi S, Kyoya S, Miyawaki T, Kidani H, Funabashi T, Nakashima H, Nakanuma Y, Ohta G, Itagaki E, Katagiri M. Cholesterol side-chain cleavage enzyme activity and cytochrome P450 content in adrenal mitochondria of a patient with congenital lipoid adrenal hyperplasia (Prader disease). Clin Chim Acta. 1977;77:301–6.
16. Hauffa BP, Miller WL, Grumbach MM, Conte FA, Kaplan SL. Congenital adrenal hyperplasia due to deficient cholesterol side-chain cleavage activity (20, 22 desmolase) in a patient treated for 18 years. Clin Endocrinol. 1985;23:481–93.
17. Chung B, Matteson KJ, Voutilainen R, Mohandas TK, Miller WL. Human cholesterol side-chain cleavage enzyme, P450scc: cDNA cloning, assignment of the gene to chromosome 15, and expression in the placenta. Proc Natl Acad Sci USA. 1986;83:8962–6.
18. Matteson KJ, Chung B, Urdea MS, Miller WL. Study of cholesterol side-chain cleavage (20, 22 desmolase) deficiency causing congenital lipoid adrenal hyperplasia using bovine-sequence P450scc oligodeoxyribonucleotide probes. Endocrinology. 1986;118:1296–305.
19. Morohashi K, Sogawa K, Omura T, Fujii-Kuriyama Y. Gene structure of human cytochrome P-450(scc), cholesterol desmolase. J Biochem. 1987;101:879–87.
20. Lin D, Gitelman SE, Saenger P, Miller WL. Normal genes for the cholesterol side chain cleavage enzyme, P450scc, in congenital lipoid adrenal hyperplasia. J Clin Invest. 1991;88:1955–62.
21. Besman MJ, Yanagibashi K, Lee TD, Kawamura M, Hall PF, Shively JE. Identification of des-(Gly-Ile)-endozepine as an effector of corticotropin-dependent adrenal steroidogenesis: Stimulation of cholesterol delivery is mediated by the peripheral benzodiazepine receptor. Proc Natl Acad Sci USA. 1989;86:4897–901.
22. Lin D, Chang YJ, Strauss JF 3rd, Miller WL. The human peripheral benzodiazepine receptor gene: cloning and characterization of alternative splicing in normal tissues and in a patient with congenital lipoid adrenal hyperplasia. Genomics. 1993;18:643–50.

23. Stocco DM, Clark BJ. Regulation of the acute production of steroids in steroidogenic cells. Endocr Rev. 1996;17:221–44.
24. Pon LA, Orme-Johnson NR. Acute stimulation of steroidogenesis in corpus luteum and adrenal cortex by peptide hormones. J Biol Chem. 1986;261:6594–9.
25. Pon LA, Hartigan JA, Orme-Johnson NR. Acute ACTH regulation of adrenal corticosteroid biosynthesis: Rapid accumulation of a phosphoprotein. J Biol Chem. 1986;261:13309–16.
26. Epstein LF, Orme-Johnson NR. Regulation of steroid hormone biosynthesis. Identification of precursors of a phosphoprotein targeted to the mitochondrion in stimulated rat adrenal cortex cells. J Biol Chem. 1991;266:19739–45.
27. Stocco DM, Sodeman TC. The 30 kDa mitochondrial proteins induced by hormone stimulation in MA-10 mouse Leydig tumor cells are processed from larger precursors. J Biol Chem. 1991;266:19731–8.
28. Clark BJ, Wells J, King SR, Stocco DM. The purification, cloning and expression of a novel luteinizing hormone-induced mitochondrial protein in MA-10 cells mouse Leydig tumor cells. Characterization of the steroidogenic acute regulatory protein (StAR). J Biol Chem. 1994;269:28314–22.
29. Saenger P, Klonari Z, Black SM, Compagnone N, Mellon SH, Fleischer A, Abrams CAL, Shackleton CHL, Miller WL. Prenatal diagnosis of congenital lipoid adrenal hyperplasia. J Clin Endocrinol Metab. 1995;80:200–5.
30. Sugawara T, Holt JA, Driscoll D, Strauss JF 3rd, Lin D, Miller WL, Patterson D, Clancy KP, Hart IM, Clark BJ, Stocco DM. Human steroidogenic acute regulatory protein (StAR): functional activity in COS-1 cells, tissue-specific expression, and mapping of the structural gene to 8p11.2 and an expressed pseudogene to chromosome 13. Proc Natl Acad Sci USA. 1995;92:4778–82.
31. Harikrishna JA, Black SM, Szklarz GD, Miller WL. Construction and function of fusion enzymes of the human cytochrome P450scc system. DNA Cell Biol. 1993;12:371–9.
32. Black SM, Harikrishna JA, Szklarz GD, Miller WL. The mitochondrial environment is required for activity of the cholesterol side-chain cleavage enzyme, cytochrome P450scc. Proc Natl Acad Sci USA. 1994;91:7247–51.
33. Toff ME, Schleyer H, Strauss JF 3rd. Metabolism of 25-hydroxycholesterol by rat luteal mitochondria and dispersed cells. Endocrinology. 1982. 111:1785–90.
34. Tee MK, Lin D, Sugawara T, Holt JA, Guiguen Y, Buckingham B, Strauss JF 3rd, Miller WL. T → A transversion 11 bp from a splice acceptor site in the gene for steroidogenic acute regulatory protein causes congenital lipoid adrenal hyperplasia. Hum Mol Genet. 1995;4:2299–305.
35. Bose HS, Sugawara T, Strauss JF 3rd, Miller WL. The pathophysiology and genetics of congenital lipoid adrenal hyperplasia. N Engl J Med. 1996;335:1870–8.
36. Miller WL, Auchus RJ. The molecular biology, biochemistry, and physiology of human steroidogenesis and its disorders. Endocr Rev. 2011;32:81–151.
37. Miller WL, Bose HS. Early steps in steroidogenesis: intracellular cholesterol trafficking. J Lipid Res. 2011;52:2111–35.
38. Porter FD, Herman GE. Malformation syndromes caused by disorders of cholesterol synthesis. J Lipid Res. 2011;52:6–34.
39. Illingworth D, Kenney TA, Orwoll ES. Adrenal function in heterozygous and homozygous hypobetalipoproteinemia. J Clin Endocrinol Metab. 1982;54:27–33.
40. Dobs AS, Schrott H, Davidson MH, Bays H, Stein EA, Kush D, Wu M, Mitchel Y, Illingworth RD. Effects of high-dose simvastatin on adrenal and gonadal steroidogenesis in men with hypercholesterolemia. Metabolism. 2000;49:1234–8.
41. Lohse P, Maas S, Sewell AC, van Diggelen OP, Seidel D. Molecular defects underlying Wolman disease appear to be more heterogeneous than those resulting in cholesteryl ester storage disease. J Lipid Res. 1999;40:221–8.
42. Vanier MT, Millat G. Niemann-Pick disease type C. Clin Genet. 2003;64:269–81.
43. Xu S, Benoff B, Liou HL, Lobel P, Stock AM. Structural basis of sterol binding by NPC2, a lysosomal protein deficient in Niemann-Pick type C2 disease. J Biol Chem. 2007;282:23525–31.

44. Kwon HJ, Abi-Mosleh L, Wang ML, Deisenhofer J, Goldstein JL, Brown MS, Infante RE. Structure of N-terminal domain of NPC1 reveals distinct subdomains for binding and transfer of cholesterol. Cell. 2009;137:1213–24.
45. Alpy F, Tomasetto C. MLN64 and MENTHO, two mediators of endosomal cholesterol transport. Biochem Soc Trans. 2006;34:343–5.
46. Zhang M, Liu P, Dwyer NK, Christenson LK, Fujimoto T, Martinez F, Comly M, Hanover JA, Blanchette-Mackie EJ, Strauss JF 3rd. MLN64 mediates mobilization of lysosomal cholesterol to steroidogenic mitochondria. J Biol Chem. 2002;277:33300–10.
47. Alpy F, Wendling C, Rio MC, Tomasetto C. MENTHO, a MLN64 homologue devoid of the START domain. J Biol Chem. 2002;277:50780–7.
48. Alpy F, Stoeckel ME, Dierich A, Escola JM, Wendling C, Chenard MP, Vanier MT, Gruenberg J, Tomasetto C, Rio MC. The steroidogenic acute regulatory protein homolog MLN64, a late endosomal cholesterol-binding protein. J Biol Chem. 2001;276:4261–9.
49. Clark BJ. The mammalian START domain protein family in lipid transport in health and disease. J Endocrinol. 2012;212:257–75.
50. Watari H, Arakane F, Moog-Lutz C, Callen CB, Tomasetto C, Gerton GL, Rio MC, Baker ME, Strauss JF 3rd. MLN64 contains a domain with homology to the steroidogenic acute regulatory protein (StAR) that stimulates steroidogenesis. Proc Natl Acad Sci USA. 1997;94:8462–7.
51. Bose HS, Whittal RM, Huang MC, Baldwin MA, Miller WL. N-218 MLN64, a protein with StAR-like steroidogenic activity, is folded and cleaved similarly to StAR. Biochemistry. 2000;39:11722–31.
52. Olvera-Sanchez S, Espinosa-Garcia MT, Monreal J, Flores-Herrera O, Martinez F. Mitochondrial heat shock protein participates in placental steroidogenesis. Placenta. 2011;32:222–9.
53. Kishida T, Kostetskii I, Zhang Z, Martinez F, Liu P, Walkley SU, Dwyer NK, Blanchette-Mackie EJ, Radice GL, Strauss JF 3rd. Targeted mutation of the MLN64 START domain causes only modest alterations in cellular sterol metabolism. J Biol Chem. 2004;279:19276–85.
54. Charman M, Kennedy BE, Osborne N, Karten B. MLN64 mediates egress of cholesterol from endosomes to mitochondria in the absence of functional Niemann-Pick Type C1 protein. J Lipid Res. 2010;51:1023–34.
55. Rodriguez-Agudo D, Ren S, Wong E, Marques D, Redford K, Gil G, Hylemon PB, Pandak WM. Intracellular cholesterol transporter StarD4 binds free cholesterol and increases cholesteryl ester formation. J Lipid Res. 2008;49:1409–19.
56. Chen Y-C, Meier RK, Zheng S, Khundmiri SJ, Tseng MT, Lederer ED, Epstein PN, Clark BJ. Steroidogenic acute regulatory (StAR)-related lipid transfer domain protein 5 (STARD5): localization and regulation in renal tubules. Am J Physiol Renal Physiol. 2009. 297:F380–8. http://ajprenal.physiology.org/content/297/2/F380.full. http://en.wikipedia.org/wiki/American_Journal_of_Physiology.
57. Letourneau D, Lorin A, Lefebvre A, Frappier V, Gaudreault F, Najmanovich R, Lavigne P, Lehoux JG. StAR-related lipid transfer domain protein 5 binds primary bile acids. J Lipid Res. 2012;53:2677–89.
58. Bose HS, Whittal RM, Ran Y, Bose M, Baker BY, Miller WL. StAR-like activity and molten globule behavior of StARD6, a male germ-line protein. Biochemistry. 2008;47:2277–88.
59. Riegelhaupt JJ, Waase MP, Garbarino J, Cruz DE, Breslow JL. Targeted disruption of steroidogenic acute regulatory protein D4 leads to modest weight reduction and minor alterations in lipid metabolism. J Lipid Res. 2010;51:1134–43.
60. Stevens VL, Xu T, Lambeth JD. Cholesterol pools in rat adrenal mitochondria: use of cholesterol oxidase to infer a complex pool structure. Endocrinology. 1992;130:1557–63.
61. Tsujishita Y, Hurley JH. Structure and lipid transport mechanism of a StAR-related domain. Nat Struct Biol. 2000;7:408–14.
62. Mathieu AP, Fleury A, Ducharme L, Lavigne P, LeHoux JG. Insights into steroidogenic acute regulatory protein (StAR)-dependent cholesterol transfer in mitochondria: evidence from molecular modeling and structure-based thermodynamics supporting the existence of partially unfolded states of StAR. J Mol Endocrinol. 2002;29:327–45.

63. Yaworsky DC, Baker BY, Bose HS, Best KB, Jensen LB, Bell JD, Baldwin MA, Miller WL. pH-dependent interactions of the carboxyl-terminal helix of steroidogenic acute regulatory protein with synthetic membranes. J Biol Chem. 2005;280:2045–54.
64. Thorsell AG, Lee WH, Persson C, Siponen MI, Nilsson M, Busam RD, Kotenyova T, Schuler H, Lethio L. Comparative structural analysis of lipid-binding START domains. PLoS ONE. 2011;6:e19521–32.
65. Romanowski MJ, Soccio RE, Breslow JL, Burley SK. Crystal structure of the Mus musculus cholesterol-regulated START protein 4 (StarD4) containing a StAR-related lipid transfer domain. Proc Natl Acad Sci USA. 2002;99:6949–54.
66. Miller WL. StAR search what we know about how the steroidogenic acute regulatory protein mediates mitochondrial cholesterol import. Mol Endocrinol. 2007;21:589–601.
67. Arakane F, Sugawara T, Nishino H, Liu Z, Holt JA, Pain D, Stocco DM, Miller WL, Strauss JF 3rd. Steroidogenic acute regulatory protein (StAR) retains activity in the absence of its mitochondrial import sequence: implications for the mechanism of StAR action. Proc Natl Acad Sci USA. 1996;93:13731–6.
68. Bose HS, Lingappa VR, Miller WL. Rapid regulation of steroidogenesis by mitochondrial protein import. Nature. 2002;417:87–91.
69. Bose HS, Whittal RM, Baldwin MA, Miller WL. The active form of the steroidogenic acute regulatory protein, StAR, appears to be a molten globule. Proc Natl Acad Sci USA. 1999;96:7250–5.
70. Baker BY, Yaworsky DC, Miller WL. A pH-dependent molten globule transition is required for activity of the steroidogenic acute regulatory protein, StAR. J Biol Chem. 2005;280:41753–60.
71. Tuckey RC, Headlam MJ, Bose HS, Miller WL. Transfer of cholesterol between phospholipid vesicles mediated by the steroidogenic acute regulatory protein (StAR). J Biol Chem. 2002;277:47123–8.
72. Baker BY, Epand RF, Epand RM, Miller WL. Cholesterol binding does not predict activity of the steroidogenic acute regulatory protein, StAR. J Biol Chem. 2007;282:10223–32.
73. Arakane F, King SR, Du Y, Kallen CB, Walsh LP, Watari H, Stocco DM, Strauss JF 3rd. Phosphorylation of steroidogenic acute regulatory protein (StAR) modulates its steroidogenic activity. J Biol Chem. 1997;272:32656–62.
74. Dyson MT, Kowalewski MP, Manna PR, Stocco DM. The differential regulation of steroidogenic acute regulatory protein-mediated steroidogenesis by type I and type II PKA in MA-10 cells. Mol Cell Endocrinol. 2009;300:94–103.
75. Shen WJ, Patel S, Natu V, Hong R, Wang J, Azhar S, Kraemer FB. Interaction of hormone-sensitive lipase with steroidogenic acute regulatory protein. J Biol Chem. 2003;278:43870–6.
76. Artemenko IP, Zhao D, Hales DB, Hales KH, Jefcoate CR. Mitochondrial processing of newly synthesized steroidogenic acute regulatory protein (StAR), but not total StAR, mediates cholesterol transfer to cytochrome P450 side chain cleavage enzyme in adrenal cells. J Biol Chem. 2001;276:46583–96.
77. Bose M, Whittal RM, Miller WL, Bose HS. Steroidogenic activity of StAR requires contact with mitochondrial VDAC1 and phosphate carrier protein. J Biol Chem. 2008;283:8837–45.
78. Rone MB, Midzak AS, Issop L, Rammouz G, Jagannathan S, Fan J, Ye X, Blonder J, Veenstra T, Papadopoulos V. Identification of a dynamic mitochondrial protein complex driving cholesterol import, trafficking, and metabolism to steroid hormones. Mol Endocrinol. 2012;26:1868–82.
79. Papadopoulos V, Miller WL. Role of mitochondria in steroidogenesis. Best Pract Res Clin Endocrinol Metab. 2012;26:771–90.
80. Papadopoulos V, Baraldi M, Guilarte TR, Knudsen TB, Lacapere JJ, Lindemann P, Norenberg MD, Nutt D, Weizman A, Zhang MR, Gavish M. New nomenclature for the peripheral-type benzodiazepine receptor based on its structure and molecular function. Trends Pharmacol Sci. 2006;27:402–9.
81. Antkiewicz-Michaluk L, Guidotti A, Krueger KE. Molecular characterization and mitochondrial density of a recognition site for peripheral-type benzodiazepine ligands. Mol Pharmacol. 1988;34:272–8.

82. Papadopoulos V, Mukhin AG, Costa E, Krueger KE. The peripheral-type benzodiazepine receptor is functionally linked to Leydig cell steroidogenesis. J Biol Chem. 1990;265:3772–9.
83. Rone MB, Liu J, Blonder J, Ye X, Veenstra TD, Young JC, Papadopoulos V. Targeting and insertion of the cholesterol-binding translocator protein into the outer mitochondrial membrane. BioChemistry. 2009;48:6909–20.
84. Lacapere JJ, Papadopoulos V. Peripheral-type benzodiazepine receptor: structure and function of a cholesterol-binding protein in steroid and bile acid biosynthesis. Steroids. 2003;68:569–85.
85. Fan J, Liu J, Culty M, Papadopoulos V. Acyl-coenzyme A binding domain containing 3 (ACBD3; PAP7; GCP60): an emerging signaling molecule. Prog Lipid Res. 2010;49:218–34.
86. Liu J, Rone MB, Papadopoulos V. Protein-protein interactions mediate mitochondrial cholesterol transport and steroid biosynthesis. J Biol Chem. 2006;281:38879–93.
87. McEnery MW, Snowman AM, Trifiletti RR, Snyder SH. Isolation of the mitochondrial benzodiazepine receptor: association with the voltage-dependent anion channel and the adenine nucleotide carrier. Proc Natl Acad Sci USA. 1992;89:3170–4.
88. Bayrhuber M, Meins T, Habeck M, Becker S, Giller K, Villinger S, Vonrhein C, Griesinger C, Zweckstetter M, Zeth K. Structure of the human voltage-dependent anion channel. Proc Natl Acad Sci USA. 2008;105:15370–5.
89. Hiller S, Garces RG, Malia TJ, Orekhov VY, Colombini M, Wagner G. Solution structure of the integral human membrane protein VDAC-1 in detergent micelles. Science. 2008;321:1206–10.
90. Mannella CA, Forte M, Colombini M. Toward the molecular structure of the mitochondrial channel, VDAC. J Bioenergy Biomembr. 1992;24:7–19.
91. Brdiczka DG, Zorov DB, Sheu SS. Mitochondrial contact sites: their role in energy metabolism and apoptosis. Biochim Biophys Acta. 2006;1762:148–63.
92. Korkhov VM, Sachse C, Short JM, Tate CG. Three-dimensional structure of TspO by electron cryomicroscopy of helical crystals. Structure. 2010;18:677–87.
93. Jamin N, Neumann JM, Ostuni MA, Vu TK, Yao ZX, Murail S, Robert JC, Giatzakis C, Papadopoulos V, Lacapere JJ. Characterization of the cholesterol recognition amino acid consensus sequence of the peripheral-type benzodiazepine receptor. Mol Endocrinol. 2005;19:588–94.
94. Midzak A, Akula N, Lecanu L, Papadopoulos V. Novel androstenetriol interacts with the mitochondrial translocator protein and controls steroidogenesis. J Biol Chem. 2011;286:9875–87.
95. Papadopoulos V, Amri H, Li H, Boujrad N, Vidic B, Garnier M. Targeted disruption of the peripheral-type benzodiazepine receptor gene inhibits steroidogenesis in the R2C Leydig tumor cell line. J Biol Chem. 1997;272:32129–35.
96. Li H, Yao Z, Degenhardt B, Teper G, Papadopoulos V. Cholesterol binding at the cholesterol recognition/interaction amino acid consensus (CRAC) of the peripheral-type benzodiazepine receptor and inhibition of steroidogenesis by an HIV TAT-CRAC peptide. Proc Natl Acad Sci USA. 2001;98:1267–72.
97. Morohaku K, Pelton SH, Daugherty DJ, Butler WR, Deng W, Selvaraj V. Translocator protein/peripheral benzodiazepine receptor is not required for steroid hormone biosynthesis Endocrinology. 2014;155:89–97. doi:10.1210/en.2013−1556
98. Miller WL. Congenital lipoid adrenal hyperplasia: the human gene knockout of the steroidogenic acute regulatory protein. J Mol Endocrinol. 1997;19:227–40.
99. Ogata T, Matsuo N, Saito M, Prader A. The testicular lesion and sexual differentiation in congenital lipoid adrenal hyperplasia. Helv Paediatr Acta. 1989;43:531–8.
100. Chen X, Baker BY, Abduljabbar MA, Miller WL. A genetic isolate of congenital lipoid adrenal hyperplasia with atypical clinical findings. J Clin Endocrinol Metab. 2005;90:835–40.
101. Bose HS, Pescovitz OH, Miller WL. Spontaneous feminization in a 46, XX female patient with congenital lipoid adrenal hyperplasia caused by a homozygous frame-shift mutation in the steroidogenic acute regulatory protein. J Clin Endocrinol Metab. 1997;82:1511–5.
102. Fujieda K, Tajima T, Nakae J, Sageshima S, Tachibana K, Suwa S, Sugawara T, Strauss JF 3rd. Spontaneous puberty in 46, XX subjects with congenital lipoid adrenal hyperplasia.

Ovarian steroidogenesis is spared to some extent despite inactivating mutations in the steroidogenic acute regulatory protein (StAR) gene. J Clin Invest. 1997;99:1265–71.
103. Voutilainen R, Miller WL. Developmental expression of genes for the steroidogenic enzymes P450scc (20, 22 desmolase), P450c17 (17α-hydroxylase/17, 20 lyase) and P450c21 (21-hydroxylase) in the human fetus. J Clin Endocrinol Metab. 1986;63:1145–50.
104. Hasegawa T, Zhao LP, Caron KM, Majdic G, Suzuki T, Shizawa S, Sasano H, Parker KL. Developmental roles of the steroidogenic acute regulatory protein (StAR) as revealed by StAR knockout mice. Mol Endocrinol. 2000;14:1462–71.
105. Nakae J, Tajima T, Sugawara T, Arakane F, Hanaki K, Hotsubo T, Igarashi N, Igarashi Y, Ishii T, Koda N, Kondo T, Kohno H, Nakagawa Y, Tachibana K, Takeshima Y, Tsubouchi K, Strauss JF 3rd, Fujieda K. Analysis of the steroidogenic acute regulatory protein (StAR) gene in Japanese patients with congential lipoid adrenal hyperplasia. Hum Mol Genet. 1997;6:571–6.
106. Kim JM, Choi JH, Lee JH, Kim GH, Lee BH, Kim HS, Shin JH, Shin CH, Kim CJ, Yu J, Lee DY, Cho WK, Suh BK, Lee JE, Chung HR, Yoo HW. High allele frequency of the p.Q258X mutation and identification of a novel mis-splicing mutation in the STAR gene in Korean patients with congenital lipoid adrenal hyperplasia. Eur J Endocrinol. 2011;165:771–8.
107. Flück CE, Maret A, Mallet D, Portrat-Doyen S, Achermann JC, Leheup B, Theintz GE, Mullis PE, Morel Y. A novel mutation L260P of the steroidogenic acute regulatory protein gene in three unrelated patients of Swiss ancestry with congenital lipoid adrenal hyperplasia. J Clin Endocrinol Metab. 2005;90:5304–8.
108. Gassner HL, Toppari J, Quinteiro Gonzalez S, Miller WL. Near-miss apparent SIDS from adrenal crisis. J Pediatr. 2004;145:178–83.
109. Baker BY, Lin L, Kim CJ, Raza J, Smith CP, Miller WL, Achermann JC. Nonclassic congenital lipoid adrenal hyperplasia: a new disorder of the steroidogenic acute regulatory protein with very late presentation and normal male genitalia. J Clin Endocrinol Metab. 2006;91:4781–5.
110. Metherell LA, Naville D, Halaby G, Begeot M, Huebner A, Nurnberg G, Nurnberg P, Green J, Tomlinson JW, Krone NP, Lin L, Racine M, Berney DM, Achermann JC, Arlt W, Clark AJ. Nonclassic lipoid congenital adrenal hyperplasia masquerading as familial glucocorticoid deficiency. J Clin Endocrinol Metab. 2009;94:3865–7381.
111. Sahakitrungruang T, Soccio RE, Lang-Muritano M, Walker JM, Achermann JC, Miller WL. Clinical, genetic, and functional characterization of four patients carrying partial loss-of-function mutations in the steroidogenic acute regulatory protein (StAR). J Clin Endocrinol Metab. 2010;95:3352–9.
112. Kirkland RT, Kirkland JL, Johnson CM, Horning MG, Librik L, Clayton GW. Congenital lipoid adrenal hyperplasia in an eight-year-old phenotypic female. J Clin Endocrinol Metab. 1973;36:488–96.
113. Linder BL, Esteban NV, Yergey AL, Winterer JC, Loriaux DL, Cassorla F. Cortisol production rate in childhood and adolescence. J Pediatr. 1990;117:892–6.
114. Esteban NV, Loughlin T, Yergey AL, Zawadzki JK, Booth JD, Winterer JC, Loriaux DL. Daily cortisol production rates in man determined by stable isotope dilution/mass spectrometry. J Clin Endocrinol Metab. 1991;72:39–45.
115. Kerrigan JR, Veldhuis JD, Leyo SA, Iranmanesh A, Rogol AD. Estimation of daily cortisol production and clearance rates in normal pubertal males by deconvolution analysis. *J*. Clin Endocrinol Metab. 1993;76:1505–10.
116. Khoury K, Barbar E, Ainmelk Y, Ouellet A, LeHoux JG. Gonadal function, first cases of pregnancy, and child delivery in a woman with lipoid congenital adrenal hyperplasia. J Clin Endocrinol Metab. 2009;94:1333–7.
117. Tajima T, Fujieda K, Kouda N, Nakae J, Miller WL. Heterozygous mutation in the cholesterol side chain cleavage enzyme (P450scc) gene in a patient with 46, XY sex reversal and adrenal insufficiency. J Clin Endocrinol Metab. 2001;86:3820–5.
118. Nebert DW, Wikvall K, Miller WL. Human cytochromes P450 in health and disease. Phil Trans R Soc B. 2013;368:20120431. http://dx.doi.org/10.1098/rstb.2012.0431.

119. Miller WL. Minireview: regulation of steroidogenesis by electron transfer. Endocrinology. 2005;146:2544–50.
120. Kuwada M, Kitajima R, Suzuki H, Horie S. Purification and properties of cytochrome P-450(SCC) from pig testis mitochondria. Biochem Biophys Res Commun. 1991;176:1501–8.
121. Mast N, Annalora AJ, Lodowski DT, Palczewski K, Stout CD, Pikuleva IA. Structural basis for three-step sequential catalysis by the cholesterol side chain cleavage enzyme CYP11A1. J Biol Chem. 2011;286:5607–13.
122. Strushkevich N, MacKenzie F, Cherkesova T, Grabovec I, Usanov S, Park HW. Structural basis for pregnenolone biosynthesis by the mitochondrial monooxygenase system. Proc Natl Acad Sci USA. 2011;108:10139–43.
123. Freeman MR, Dobritsa A, Gaines P, Segraves WA, Carlson JR. The *dare* gene: steroid hormone production, olfactory behavior, and neural degeneration in Drosophila. Development. 1999;126:4591–602.
124. Yang X, Iwamoto K, Wang M, Artwohl J, Mason JI, Pang S. Inherited congenital adrenal hyperplasia in the rabbit is caused by a deletion in the gene encoding cytochrome P450 cholesterol side-chain cleavage enzyme. Endocrinology. 1993;132:1977–82.
125. Hu MC, Hsu NC, El Hadj NB, Pai CI, Chu HP, Wang CKL, Chung BC. Steroid deficiency syndromes in mice with targeted disruption of Cyp11a1. Mol Endocrinol. 2002;16:1943–50.
126. Miller WL. Why nobody has P450scc (20, 22 desmolase) deficiency. (Letter to the Editor). J Clin Endocrinol Metab. 1998;83:1399–400.
127. Tee MK, Abramsohn M, Loewenthal N, Harris M, Siwach S, Kaplinsky A, Markus B, Birk O, Sheffield VC, Pavari R, Hershkovitz E, Miller WL. Varied clinical presentations with mutations in *CYP11A1* encoding the cholesterol side chain cleavage enzyme, P450scc. J Clin Endocrinol Metab. 2013;98:713–20.
128. Rubtsov P, Karmanov M, Sverdlova P, Spirin P, Tiulpakov A. A novel homozygous mutation in CYP11A1 gene is associated with late onset adrenal insufficiency and hypospadias in a 46 XY patient. J Clin Endocrinol Metab. 2009;94:936–9.
129. Sahakitrungruang T, Tee MK, Blackett PR, Miller WL. Partial defect in the cholesterol side-chain cleavage enzyme P450scc (CYP11A1) resembling nonclassic congenital lipoid adrenal hyperplasia. J Clin Endocrinol Metab. 2011;96:792–8.
130. Gucev Z, Tee MK, Chitayat D, Wherrett D, Miller WL. Distinguishing deficiencies of StAR and P450scc causing neonatal adrenal failure. J Pediatr. 2013;162:819–22.
131. Schimmer BP, White PC. Steroidogenic factor 1: Its roles in differentiation, development and disease. Mol Endocrinol. 2010;24:1322–37.
132. Achermann JC, Ito M, Hindmarsh PC, Jameson JL. A mutation in the gene encoding steroidogenic factor-1 causes XY sex reversal and adrenal failure in humans. Nat Genet. 1999;22:125–6.
133. Camats N, Pandey AV, Fernandez-Cancio M, Andaluz P, Janner M, Toran N, Moreno F, Bereket A, Akcay T, Garcia-Garcia E, Munoz MT, Gracia R, Nistal M, Castano L, Mullis PE, Carrascosa A, Audi L, Flück CE. Ten novel mutations in the NR5A1 gene cause disordered sex development in 46, XY and ovarian insufficiency in 46, XX individuals. J Clin Endocrinol Metab. 2012;97:E1294–306.

Chapter 5
Steroidogenic Acute Regulatory Protein (StAR) and Atherogenesis

Annette Graham, Faye Borthwick and Janice Taylor

Abstract Atherosclerosis, the primary cause of coronary heart disease, is characterised by a low-grade unresolved inflammation associated with accumulation of cholesterol and cholesteryl-ester laden macrophages within the intima of the vessel wall. Steroidogenic acute regulatory protein (StAR/STARD1) is endogenously expressed, and regulated, in a number of vascular tissues, including endothelial cells and monocyte-macrophages where it is thought to traffic cholesterol from the outer to the inner mitochondrial membrane, determining the rate at which substrate is supplied to sterol 27-hydroxylase (CYP27A1). The CYP27A1 enzyme converts cholesterol to oxysterol derivatives, which act as activating ligands for nuclear Liver X Receptors, master regulators of lipid metabolism and inflammatory responses. Forced overexpression of StAR/STARD1 in macrophages and endothelial cells increases the cholesterol efflux process mediated by adenosine triphosphate (ATP)-binding cassette transporters (ABCA1/G1) and apolipoprotein acceptors, and inhibits nuclear factor-κB signalling, resulting in repression of an array of inflammatory genes. Thus, StAR/STARD1 may represent a novel target for treatment of atherosclerosis and coronary heart disease.

Abbreviations

ACAT	Acyl CoA: Cholesterol AcylTransferase
ACBD	Acyl CoA Binding Domain Protein
ANT	Adenine Nucleotide Transporter
Apo	Apolipoprotein
ASTAD3A	ATPase Family AAA Domain-containing Protein 3A
CD	Cluster of Differentiation
CETP	Cholesteryl Ester Transfer Protein
ChREBP	Binding Protein
CTX	Cerebrotendinous Xanthomatosis

A. Graham, Ph.D. (✉) · J. Taylor, Ph.D.
Department of Life Sciences, School of Health and Life Sciences,
Glasgow Caledonian University, Glasgow, UK
e-mail: Ann.Graham@gcu.ac.uk

F. Borthwick, Ph.D.
Metabolic and Cardiovascular Diseases Laboratory, Alberta Diabetes Institute,
University of Alberta, Edmonton, AB, Canada

B. J. Clark, D. M. Stocco (eds.), *Cholesterol Transporters of the START Domain Protein Family in Health and Disease,* DOI 10.1007/978-1-4939-1112-7_5,

COX	Cyclooxygenase
CXCL16	Chemokine (C-X-C motif) Ligand 16
CYP11A1	Cytochrome P450 Side Chain Cleavage Enzyme
CYP27A1	Sterol 27-Hydroxylase
CYP51A1	Lanosterol 14α-methylase
DR	Direct Repeat
ER	Estrogen Receptor
FAS	Fatty Acid Synthase
GM-CSF	Granulocyte Macrophage Colony Stimulating Factor
HDL	High-density Lipoprotein
HDL-C	HDL-cholesterol
HMG CoA Reductase	3-Hydroxy 3-MethylGlutaryl CoA Reductase
IFN	Interferon
IL	Interleukin
iNOS	Inducible Nitric Oxide Synthase
LCAT	Lecithin:Cholesterol AcylTransferase
LDL	Low Density Lipoprotein
LOX-1	Lectin-like Oxidized LDL Receptor
LPS	Lipopolysaccharide
LXR	Liver X Receptor
LXRE	Response Element
MCP	Monocyte Chemotactic Protein
M-CSF	Macrophage Colony Stimulating Factor
MLN64; STARD3	Metastatic Lymph Node 64
NF-κB	Nuclear Factor Kappa B
NCoR	Nuclear Receptor Co-Repressor
PKA	Protein Kinase A
PLTP	Phospholipid Transfer Protein
PPAR	Peroxisome Proliferator Activated Receptor
RXR	Retinoid X Receptor
SCD	Stearoyl CoA Desaturase
SERM	Selective Estrogen Receptor Modulator
SMRT	Silencing Mediator of Retinoid and Thyroid receptors
SR-AI/AII	Scavenger Receptor AI/AII
SRE	Sterol Regulatory Element
SREBP	Binding Protein
StAR; STARD1	Steroidogenic Acute Regulatory protein
STARD3	StAR-related Lipid Transfer Domain 3 Protein
TGF	Transforming Growth Factor
TNF	Tumour Necrosis Factor
TSPO	18 kDa Translocator Protein
VCAM	Vascular Cell Adhesion Molecule
VDAC	Voltage Dependent Anion Channel

Introduction

My introduction to the biological significance of steroidogenic acute regulatory protein (StAR) came in the late 1990s from a seminar given at the Hampstead campus of Royal Free and University College Medical School, London by Professor Douglas M. Stocco. Professor Stocco was there at the invitation of his fellow endocrinologist, and my Head of Department, Professor Brian Cooke, to discuss the role of StAR in acute hormone-stimulated regulation of steroidogenesis. A fascinating seminar ensued, covering deletion studies elucidating the (non-essential) role of the N-terminal targeting sequence of StAR in mitochondrial cholesterol trafficking and discussing the unequivocal evidence supporting mutations in StAR rather than the peripheral benzodiazepine receptor (PBR; 18 kDa translocator protein, TSPO) as the definitive culprit in congenital lipoid hyperplasia. My research interests, at that time, lay in the mechanisms by which cholesterol could be modified to oxysterols by vascular cells. Reflecting the role of StAR in steroidogenesis, the rate-limiting factor for modification of cholesterol by sterol 27-hydroxylase (CYP27A1) within the mitochondrial matrix, appeared to be the movement of cholesterol from the outer to the inner mitochondrial membrane (*discussed below*). Following discussions on the potential role and/or possible expression of StAR in non-steroidogenic tissues such as macrophages, Professor Stocco very kindly agreed to supply me with a full-length clone, and antibody, for murine StAR to test the hypothesis that overexpression of StAR could increase the flux of cholesterol to oxysterol in these cells. This anti-atherogenic concept was funded by the British Heart Foundation and, after a career break occasioned by the birth of my twins, my group finally proved the validity of this hypothesis in an article published in Cardiovascular Research entitled 'Over-expression of steroidogenic acute regulatory protein (StAR) increases macrophage cholesterol efflux to apolipoprotein AI' in 2010.

This chapter focuses on the anti-atherogenic concept for StAR within the context of recent research. Following current nomenclature, StAR is designated STARD1, however, throughout this chapter, I will use the original classical name, StAR, to refer to the gene and protein.

Background

Coronary heart disease is the leading cause of death worldwide, causing the death of 3.8 million men and 3.4 million women each year, according to recent figures from the World Health Organisation, and despite improvements in treatment strategies and in survival rates, 1 in 4 men and 1 in 3 women, still die within 1 year following a myocardial infarction. Coronary heart disease is fuelled by genetic and environmental factors, including high levels of serum cholesterol which cause one-third of all deaths from cardiovascular disease worldwide. The primary cause of coronary heart disease is atherosclerosis, a slow progressing disease of large- and medium-sized arteries, characterised by a chronic and unresolved inflammatory response at sites of perturbed laminar blood flow [1–3]. The process of atherogenesis is complex, involving interactions between lipoproteins, leucocytes (monocytes, neutrophils, lymphocytes and possibly circulating stem cells) and platelets from the bloodstream, with endothelial, smooth muscle, mast and dendritic cells within the artery wall. Following activation or dysregulation of the endothelium lining the vessel, lipoproteins, such as low-density lipoproteins (LDL) can accumulate within the intima and become modified via oxidation, triggering an immune response responsible for attracting and recruiting leucocytes to this site. Once there, monocytes differentiate into macrophages, and polarise to generate a heterogeneous phenotypic pool, depending upon the stimuli present in the microenvironment. Pro-inflammatory (M1) and anti-inflammatory (M2) macrophages represent distinct extremes of this phenotypic spectrum, which is influenced by differentiation factors, such as granulocyte macrophage colony stimulating factor (GM-CSF; M1) and M-CSF (M2), by priming T-cell derived cytokines such as interferon (IFN)-g and interleukin (IL)-13 (M1) and IL-4 and IL-10 (M2), and by the presence of oxidized LDL [3].

Loss of effective macrophage cholesterol homeostasis lies at the heart of early (fatty streak), developing and unstable atherosclerotic lesions [4–6]. Macrophage 'foam cells' become laden with cholesterol and cholesteryl ester, in part from the uptake of apoptotic cells, and in part via the unregulated uptake of modified lipoproteins, such as oxidized LDL, by scavenger receptors such as cluster of differentiation (CD)36, CD68, lectin-like oxidized LDL receptor 1 (LOX-1) and SR-AI/AII, forming early 'fatty streak' lesions [4–7]. During this early phase, macrophage recognition of modified lipoproteins, and phagocytosis of cellular debris and necrotic/apoptosed cells, may help to protect against atheroma progression; however, in more advanced lesions, cholesterol-laden macrophages, by releasing inflammatory cytokines and matrix metalloproteinases, contribute to chronic unresolved inflammation, leading to acceleration of the disease process and acute thrombotic states, such as myocardial infarction or stroke [7].

Removal of cholesterol, from macrophage 'foam cells' within the artery wall may therefore be a key factor in achieving successful regression and stabilisation of atheroma. This goal can be achieved, to varying degrees of efficacy, by the reverse cholesterol transport pathway mediated by high-density lipoproteins (HDL), as evidenced by epidemiological studies in humans, and in genetically modified mice in which components of this pathway have been overexpressed or deleted [8–12]. In the Framingham Study, risk for coronary artery disease (CAD) rose sharply as levels of HDL fell below 40 mg/dl. Plasma levels of HDL-cholesterol (HDL-C) were inversely predictive of coronary heart disease in the Lipid Research Clinics Primary Prevention Trial [10]. However, HDL particles not only aid removal of cholesterol from peripheral tissues, but also possess antioxidant properties [8–12], and can counteract the chronic inflammation [8–12], proliferation of haematopoetic stem cells [13] and leucocytosis which promote atherosclerosis [14]. Intriguingly, it is clear that in patients with systemic inflammation, coronary heart disease, chronic renal disease, and diabetes, the protective, anti-atherogenic properties of HDL are lost, or the particles are transformed into those with pro-atherogenic potential [12]. These findings indicate that it is not just the level, but the quality and function of the HDL particles that are important.

Some, but not all, of the beneficial anti-atherogenic effects associated with HDL are mediated by interactions with adenosine triphosphate (ATP)-binding cassette (ABC) transporters, ABCA1, ABCG1 and ABCG4, which promote cellular cholesterol efflux, the first step in the reverse cholesterol transport process [9]. While ABCA1 promotes the efflux of cholesterol and phospholipids to lipid-poor acceptors, such as Apolipoprotein (Apo)A-I and ApoE [15], ABCG1 and ABCG4 promote efflux of cholesterol, oxysterols and desmosterol to HDL [16]. Thus, these transporters work in concert to generate HDL, which subsequently mature within the reverse cholesterol transport pathway in the bloodstream, via interactions with lecithin cholesterol acyltransferase (LCAT), phospholipid transfer protein (PLTP) and cholesteryl ester transfer protein (CETP), enabling the transfer of excess cholesterol from peripheral tissues to the liver for excretion in the form of bile and bile acids [9, 15, 16].

Liver X Receptors: Master Regulators of Lipid Metabolism and Inflammatory Responses

Expression of ABCA1, ABCG1 and ABCG4 lie under the control of nuclear Liver X receptors (LXRα/β), which are pivotal in orchestrating appropriate cellular responses to elevated sterol content [17–19]; both synthetic and oxysterol LXR ligands potently up-regulate ABCA1, ABCG1 and ABCG4 gene expression in human monocytes and macrophages. Liver X receptors form heterodimers with retinoid X receptors (RXRs), bound to an imperfect direct repeat of the nuclear receptor half-site, TGACCT, separated by four bases (direct repeat (DR)-4) [17]. Further, cholesterol biosynthesis can be repressed by LXRα, via novel negative LXR deoxyribonucleic acid (DNA) response elements within the promoter of squalene synthase and lanosterol 14α-methylase (CYP51A1) [20], while uptake of LDL is blocked by transcriptional induction of the E3 ubiquitin ligase Idol (inducible degrader of the LDL receptor) [21]. These events, together with sequestration of sterol regulatory element binding proteins (SREBPs) at the endoplasmic reticulum, and storage of excess cholesterol via Acyl CoA: Cholesterol Acyltransferase (ACAT) as cytosolic droplets of cholesteryl-ester, form a coordinated homeostatic response, preventing excess sterol accumulation [22, 23]. Cholesterol deposition in peripheral tissues and accelerated atherosclerosis ensue, following deletion of the gene encoding LXRα (*NR1H3*) in $ApoE^{-/-}$ mice, while loss of both LXRα and LXRβ isoforms elicited lipid accumulation within foam cells in the aortic root, despite the absence of an atherogenic stimulus [reviewed in 22].

Further, LXRs transrepress genes involved in the inflammatory response, via a mechanism involving nuclear receptor co-repressor (NCoR), silencing mediator of retinoid and thyroid receptors (SMRT) and inhibition of nuclear factor κB

(NF-κB) signalling [24]. Thus, activation of LXR results in repression of an array of inflammatory genes in macrophages, stimulated by exposure to bacteria, lipopolysaccharide (LPS), tumour necrosis factor (TNF-α) or interleukin-1β (IL-1β) [reviewed in 25]. These genes include inducible nitric oxide synthase (iNOS), cyclooxygenase-2 (COX-2), interleukin-6 (IL-6) and IL-1β, as well as chemokines implicated in monocyte recruitment to the vessel wall, such as monocyte chemoattractant protein-1 (MCP-1) and MCP-3. Importantly, both LXRα and LXRβ appear to possess equivalent anti-inflammatory activity, as LXR ligands can repress these genes in macrophages derived from either LXRα or LXRβ null mice, but not from macrophages null for both isoforms of this nuclear receptor [26].

Finally, while it is clear from the above that LXRs integrate cholesterol metabolism and inflammatory responses, it is evident that these transcription factors also play a significant role in the control of hepatic fatty acid metabolism. Disruption of the gene encoding LXRα, leads to loss of expression of Sterol Regulatory Element Binding protein-1c (SREBP-1c), fatty acid synthase (FAS) and stearoyl CoA desaturase-1 (SCD-1) in mice [25]. Certainly, treatment with synthetic LXR ligands triggers increases in hepatic and plasma triglyceride levels, posing a challenge for the development of these molecules as anti-atherogenic therapeutic agents [27]. LXR also directly activates the expression of the gene encoding carbohydrate response-element binding protein (ChREBP), the other major transcriptional activator of hepatic lipogenesis [28]. Thus, LXR, SREBP-1c and ChREBP work in combination to increase the expression of genes needed for hepatic lipogenesis and triglyceride biosynthesis.

Macrophage Generation of Oxysterol Ligands for Liver X Receptors

Potential sources of oxysterol ligands for LXRs within developing atherosclerotic lesions could derive from uptake of oxidized LDL [22, 23], from the cholesterol biosynthetic pathway [20], or via cholesterol modification by sterol 27-hydroxylase (CYP27A1) [29–32] at the interface between the inner mitochondrial membrane and the matrix. However, the major oxysterols in oxidized LDL, 7-keto and 7-β-hydroxycholesterol do not bind LXRα [17–19], and formation of 24(S), 25-epoxycholesterol via the cholesterol biosynthetic pathway is profoundly repressed in cholesterol-enriched macrophage 'foam' cells [20]. By contrast, CYP27A1 can produce a number of oxysterol ligands for LXRα [31], the role of 27-hydroxycholesterol as an LXR ligand has been confirmed in vivo [32], and production of 27-hydroxycholesterol has been directly linked with LXRα activation, and induction of ABCA1 mRNA and protein in macrophages [33, 34], endothelial cells [35], enterocytes [36, 37] and neurons [38]. Loss of CYP27A1 is associated with loss of ABCA1 induction in response to cholesterol loading in human fibroblasts [39]; equally, diminished 27-hydroxycholesterol production occurs in murine macrophages depleted of Niemann-Pick C protein [40], and human fibroblasts with cholesteryl ester storage disease [41], both conditions associated with impaired up-regulation of ABCA1 or ABCG1 and reduced cholesterol efflux.

The gene encoding human CYP27A1 is located on chromosome 2q35, producing a cytochrome P450 enyzme consisting of a 498-amino acid mature protein, localised by a 33 amino acid mitochondrial signal sequence. The protein has established adrenodoxin (residues 351–365) and heme-binding sites (residues 435–464), which interact with cofactors, adrenodoxin and adrenodoxin reductase to mediate electron transfer from NAPDH and facilitate the oxidation of the cholesterol side chain [reviewed in 42]. The autosomal recessive lipid storage disorder, cerebrotendinous xanthomatosis (CTX), is caused by mutated forms of CYP27A1 [42–44]. While the familial phenotype of CTX is complex, and cannot be reproduced by genetic deletion in mice [44], patients accumulate cholesterol and cholestanol in tissues, suffer progressive neurological deterioration and, as a secondary complication in 10 % of individuals, accelerated atherosclerosis [43]. Importantly, delivery of cholesterol to the inner mitochondrial membrane, via the mitochondrial cholesterol-trafficking complex, has been proven to be the rate-limiting factor controlling the activity of CYP27A1 [45, 46].

Supply of Cholesterol to CYP27A1: A Role for StAR

One current model by which cholesterol is transferred from the outer to the cholesterol-poor inner mitochondrial membrane, involves a basal complex, forming contact sites between these membranes, composed of the 18 kDa translocator protein (TSPO), VDAC (voltage-dependent anion channel) and ANT (adenine nucleotide transporter) [reviewed in 47, 48]. In steroidogenic tissues subjected to acute hormonal stimulation, a ‘transduceosome’ complex forms, involving the recruitment of the regulatory subunits of protein kinase A (PKA-R1α) and acyl CoA binding domain proteins-1 and -3. Elevated levels of cyclic adenosine monophosphate (cAMP) then release PKA catalytic subunits to phosphorylate StAR at the outer mitochondrial membrane; import of both StAR and cholesterol into the inner mitochondrial membrane and matrix facilitate both the proteolytic processing of StAR to its 30 kDa form, and the concomitant conversion of cholesterol into pregnenolone [47–49]. More recently, Rone et al. [50] described a dynamic 800 kDa bioactive protein complex in steroidogenic cells, spanning the outer and inner mitochondrial membranes, consisting of TSPO, VDAC, cytochrome P450 side chain cleavage enzyme (CYP11A1), ATPase family AAA domain-containing protein 3 A (ASTAD3A) and optic atrophy type 1 proteins, but not ANT [50]. Addition of StAR mobilised the cholesterol bound to the 800 kDa complex, leading to increased steroid formation by CYP11A1. Whether the same proteins are involved in creating a suitable microenvironment for cholesterol import, trafficking and metabolism by CYP27A1 is not known.

It is clear, however, that the expression and functional role(s) of StAR in non-steroidogenic tissues are concepts currently gaining ground [reviewed in 51]. The expression of StAR in endothelial cells was first demonstrated by Ning et al. [52] in venules of human connective tissue, intima of human aorta, primary cultures of rat heart microvascular endothelial cells and the murine brain microvascular endothelial cell line (bEnd.3) [52]. Castillo and co-workers also verified the presence of immunoreactive StAR in endothelial cells in rodent aortae [53]. Ma et al. [54] discovered the expression of StAR in primary peritoneal macrophages isolated from ApoE$^{-/-}$ and C57BL/6 J mice, and in the RAW 264.7 (Abelson murine leukaemia virus-induced tumor) murine macrophage cell line [54]. Borthwick et al. [55] confirmed the expression of StAR in human THP-1 monocyte-macrophages, and in human peripheral blood monocytes and human heart aorta [55, 56]. It should be recognised, however, that the expression of StAR in vascular cells is far lower (10^{-2} to 10^{-4}) than those found in adrenal or gonadal tissues [51].

Regulation of Expression of StAR in Vascular Cells and Tissues

Free cholesterol, LDL-cholesterol and 25-hydroxycholesterol have been shown to increase levels of StAR mRNA and protein in the bEnd.3 endothelial cell line in a time (0–24 h) and dose-dependent (0–20 μg ml^{-1}) manner [52], findings consistent with a role for StAR in the atheroprotective cholesterol efflux pathway. Inflammatory cytokines, particularly those which either prime the generation, or are produced as effectors from, 'classically activated' (M1) macrophages, can trigger decreases in StAR expression in murine RAW 264.7 macrophages [54]. In particular, tumour necrosis factor-α (TNF-α; *Tnfsf2*) and interferon-γ (IFN-γ) exert dose (1–100 ng ml^{-1}) and time (0–12 h) dependent decreases in StAR mRNA and protein, while the anti-inflammatory cytokine, transforming growth factor-β (TGF-β; 100 ng ml^{-1}) increases the expression of StAR messenger ribonucleic acid (mRNA) and protein [54]. Indeed, loss of StAR might conceivably form part of the mechanism by which ABCA1 and ApoE are reduced in IFN-γ primed M1 macrophages, resulting in increased storage of cholesteryl ester [2, 3]. No data is available on the regulation of StAR in alternatively activated (M2a) or regulatory (M2c) macrophage subsets.

In 2009, the differential regulation of StAR, and its STARD1 sub-family member STARD3 (Metastatic Lymph Node 64; MLN64) was described in human monocyte-derived (THP-1) macrophages [55]. Analysis of the upstream promoter regions (−3 kb) of both genes indicated numerous putative sites for lipid responsive transcription factors, which for StAR included SREBP, LXR and peroxisome proliferator activated receptor (PPAR) response elements [55]. The differentiation of THP-1 monocytes into macrophages, driven by the addition of phorbol ester, was accompanied by increasing levels of phospholipids, triacylglycerol and cholesterol; levels of StAR protein were positively and significantly correlated with increases in macrophage cholesterol mass, and with ABCA1 protein. Conversely, acute depletion of cholesterol from differentiated macrophages, using methyl β-cyclodextrin or HMG-CoA reductase inhibitor, lovastatin, reduced steady state levels of StAR mRNA and/or protein [55]. Gene expression of StAR was induced by agonists of LXR (10 µM T091317), PPARγ (10 µM GW1929) and the retinoid X receptor (5 µM 9-*cis*-retinoic acid), but not by agonists of PPARα (10 µM GW7647) or PPARδ (10 µM GW0742). These data suggest StAR is an integral 'feed-forward' element of the macrophage cholesterol efflux pathway, activated by increases in sterol content: LXR/RXR heterodimers increases StAR expression, enhancing the mitochondrial transfer of cholesterol to CYP27A1, and amplifying expression of ABCA1 and/or ABCG1 [55].

Intriguingly, treatment of macrophages with the novel hypocholesterolaemic agent, LY295427 (20 µM), which releases active SREBPs from the endoplasmic reticulum, also induced the expression of StAR, implying that one, or more, of the putative SREs within the StAR promoter may be functional in these cells. Christenson et al. [57] have reported that both SREBP-1a and SREBP-1c stimulate human StAR promoter activity in adrenal cells; in macrophages, this may represent a putative feedback control mechanism, whereby increased cholesterol biosynthesis generates oxysterol ligands for LXR [55], preventing excess cholesterol accumulating within the cells. Finally, levels of StAR mRNA were significantly repressed in macrophages loaded with excess cholesterol, derived from either oxidised or acetylated LDL, implying that loss of this protein may contribute to the aetiology of atherosclerosis [55].

Functions of StAR in Vascular Cells and Tissues

Few studies have examined the functional impact of endogenous StAR in vascular cells and tissues. Conditional knockout studies in murine models of atheroma, designed to examine the impact of altered vascular StAR expression on the pathogenesis of atherosclerosis, have not been performed to date. However, combined niacin and chromium treatment, which increases the level of StAR in endothelial cells [58], improved vascular endothelial function in hyperlipidaemic rats [59]. Administration of chromium chloride (250 µg kg^{-1}) and niacin (100 mg kg^{-1}) per day by gavage to rats fed a high-fat diet improved the lipoprotein profile, down-regulated the signalling pathway mediated by the interaction of oxidized LDL with the lectin-like oxidized LDL receptor and increased the output of the vasodilator nitric oxide [59]. Clearly, however, this study can only suggest a role for StAR in regulation of endothelial cell function as niacin exerts multiple effects on lipid metabolism and cardiovascular disease [60]. Thus, the majority of studies examining the function of StAR in vascular cells have employed forced overexpression of this protein, using either plasmid or viral vectors, with some studies utilising cAMP analogues or phosphodiesterase inhibitors to stimulate PKA-dependent phosphorylation of StAR to promote the full activity of this protein [61].

Links between StAR and cholesterol homeostasis in endothelial cells were suggested by Ning et al. [62]. Overexpression (50-fold) of StAR protein, achieved by adenoviral expression (48 h), increased the levels of ABCA1 and ABCG1 mRNA and protein by 2- to 3-fold [62] in brain microvascular endothelial cells. Further, overexpression of StAR in primary rat aortic endothelial cells, challenged with palmitic acid (200 μM), blocked the activation of NF-κB, and decreased the mRNA and/or protein levels of IL-1β, TNFα, IL-6 and VCAM-1 [63]. Only a modest decrease (20–30 %) in each of these parameters was observed. Overexpression of StAR also sustained the ability of these cells to activate the protein kinase B pathway, which is linked to maintenance of endothelial nitric oxide synthase activity and nitric oxide production [63]. Treatment with small interfering RNA (siRNA) targeted against endogenous StAR before challenge with palmitic acid reversed these effects and could also be reproduced by inhibitors of fatty acid synthase (cerulenin) and cholesterol biosynthesis (lovastatin) [63].

Overexpression of StAR (130-fold) in THP-1 monocyte-derived macrophages decreased the neutral lipid levels, and the total cholesterol mass, by about 50 %, following exposure to oxidized LDL (50 μg ml^{-1} for 24 h) [64]. mRNA and protein levels of CYP27A1, LXRα, PPARγ and ABCG1 were all increased by around 2 to 3-fold after 24 to 48 h in this study, leaving levels of SREBP-1/2 and ABCA1 unchanged; these changes in expression were blocked by treatment with the LXR antagonist, 4, 4-diisothiocyanatostilbene-2, 2-disulfonic acid DIDS, confirming the key role of LXRα activation in these events. Equally, StAR overexpression in macrophages that were challenged with lipopolysaccharide (LPS) decreased the levels of interleukin-1β and TNFα, but not IL-6, secreted into the medium [64].

Taylor et al. [65] generated RAW 264.7 murine macrophages that stably overexpressed more modest levels of StAR (2.5-fold), thus avoiding any potential promiscuous effects associated with expression of this protein at higher levels [65]. A mutated, inactive form of StAR (R181L) was also expressed at the same level. In good agreement with Ning et al. [64], StAR overexpression activated and induced the expression of LXRα; however, in this study, the effects of StAR were evident only in the presence of cAMP analogue (dibutyryl cAMP; 0.3 mM), indicating normal regulation of StAR activity. Importantly, the activation of the LXRE reporter plasmid was blocked by LXRα antagonist, geranylgeranylpyrophosphate (GGPP; 10 μM) and by the sterol 27-hydroxylase inhibitor, GW273297 × (1 μM) in StAR-overexpressing but not control macrophages. Changes in expression of an array of genes involved in cholesterol metabolism and lipoprotein signalling were induced by StAR overexpression [65]. Some were genes known to be down-regulated by oxysterol mediated sequestration of SREBPs at the endoplasmic reticulum (*Fdps, Hmgcr, Mvk, Ldlr*). By contrast, ABCA1 mRNA and protein expression were markedly increased, reflecting oxysterol activation of LXRα, which translated into increased efflux of cholesterol to ApoAI in StAR overexpressing macrophages. Again, increases in cholesterol efflux caused by StAR overexpression were shown to be dependent upon the activities of both LXR and sterol 27-hydroxylase [65].

Activation of LXRα is also consistent with the decreased expression of the inflammatory cytokine, TNFα observed in StAR overexpressing macrophages[1] How-

[1] Taylor Janice, Allen Anne-Marie, Borthwick Faye, Graham Annette (2009) Q-PCR analysis of StAR-induced genes involved in lipid and lipoprotein metabolism. Glasgow Caledonian University.

ever, more surprising findings emerged from microarray analyses, such as the down-regulation of *Abca2,* previously reported to be regulated in a similar way as Abca1, with which it shares a high degree of structural homology [66]. Recent data suggest that loss of Abca2, which is thought to sequester LDL cholesterol within lysosomes, modulates intracellular cholesterol deposition, increasing macrophage apoptosis and decreasing early atherosclerosis in LDL receptor$^{-/-}$ mice [67]. Equally, StAR overexpression reduced levels of the *Sorl1* gene, which encodes LR11, a member of the LDL receptor family [68] that is highly expressed in atheromatous lesions, particularly in intimal, but not medial smooth muscle cells. Increased circulating levels of soluble LR11 correlate with intima-media thickness in dyslipidemic patients, and in individuals with stable and acute coronary disease [68]. Finally, increased expression of the gene encoding chemokine (C-X-C motif) ligand 16 (CXCL16) was noted in StAR overexpressing macrophages [65]; this multi-functional protein acts as a chemokine, a scavenger receptor and an adhesion molecule, and is highly expressed in atherosclerotic lesions in murine models of atheroma; to date, however, genetic deletion studies of CXCL16 in mice have provided conflicting results [69].

Adenoviral overexpression of StAR protein (Ad-StAR) also decreased the levels of total, early and end-stage apoptosis in THP-1 monocyte-macrophages, and prevented the induction of total and end-stage apoptosis in cells exposed to oxidized LDL (25 μg ml^{-1}). In this study, by Bai et al. [70], the level of endogenous (non-triggered) apoptosis appear to be quite high, approaching 30 %, and similar to that seen following exposure to oxidized LDL [70]. Chromatin condensation, margination, DNA laddering, apoptotic body and lipid droplet formation were all decreased in cells transfected with Ad-StAR. Further, levels of mRNA and protein for the pro-apoptotic factors, Caspase-3 and Bax, were decreased consistently by StAR overexpression, in the presence or absence of oxidised LDL, whereas levels of mRNA and protein encoding the anti-apoptotic protein, Bcl-2, increased significantly under the same conditions [70].

While numerous studies over the last decade, have carefully detailed the consequences of loss of StAR function in steroidogenic tissues [71–73], only one study to date has attempted to assess the effect of StAR overexpression in vivo in the context of atherogenesis [74]. In 2009, the Yin group [74] administered adenoviral vectors (1×10^{11} virus particles) encoding StAR (Ad-CMV-StAR) or the same vector expressing the enhanced green fluorescence protein (Ad-CMV-EGFP) via tail vein injection, to ApoE$^{-/-}$ mice aged 6 months that were fed a normal rodent chow diet. Six days after injection, StAR expression was measured in heart, coronary artery and liver, and blood, aorta and liver lipids were assessed [74]. As expected, given this route of administration, the levels of StAR were substantively increased in the liver, with marginal increases in StAR expression reported for heart and aortic tissues. Serum cholesterol and triglycerides decreased by 20–30 % in mice receiving the Ad-CMV-StAR, whereas levels of HDL-cholesterol rose by about 40 % [74]. Hepatic cholesterol and triglyceride levels fell by 25 and 56 %, respectively, in mice treated with Ad-CMV-StAR compared with control mice. Despite the low levels of StAR observed in coronary aortae, the area of lipid accumulation at this site decreased substantively, from around 20 % in control mice to around 4 % in Ad-CMV-StAR treated mice [74].

Finally, several studies have reported increased expression of StAR in failing heart, and myocardial infarction [75, 76]. Most recently, Anuka et al. [77] demonstrated increased expression of StAR mRNA and protein following generation of permanent myocardial infarction by ligation of the left anterior descending coronary artery in female BALB/c mice. However, no increases in the other enzymes required for steroid hormone synthesis, such as CYP11A1, were noted, suggesting an alternative function for StAR [77]. Notably, expression of StAR was found to be restricted to interstitial heart fibroblasts, rather than cardiomyocytes. Moreover, using neonatal heart cells, and human heart fibroblasts, the authors showed induction of StAR protein following treatment with stressors such as hydrogen peroxide (10 μM; 24 h) and staurosporine (0.1 μM; 3 h), suggesting it may function as a pro-survival factor. Indeed, the presence of StAR inhibited activation of the intrinsic apoptotic pathway, reducing the release of cytochrome c and preventing activation of BAX (Bcl2-associated X protein), and blocking fragmentation of nuclear DNA [77]. Whether the pro-survival function of StAR is related to its role as a cholesterol transporter is not clear; however, the authors speculate that the presence of StAR may facilitate proliferation and differentiation of heart fibroblasts which can help to repair and remodel heart tissue at sites of myocardial infarction.

Summary

Targeting StAR as a Therapeutic for Vascular Disease: More Questions than Answers The studies described to date suggest that increasing the levels of StAR in vascular tissues can exert anti-atherogenic effects, reducing inflammation and cholesterol accumulation (Fig. 5.1). However, there are some key questions which remain unanswered with regard to the impact of StAR in a vascular context. One is whether this protein is expressed endogenously at levels high enough to impact on vascular function or disease progression. Targeted deletion and/or overexpression studies, affecting StAR expression specifically in endothelial cells, or monocyte-macrophages, and studying the pathogenesis of atherosclerosis in murine models of this disease are clearly warranted. Alternatively, other members of the START family of lipid trafficking proteins could prove better therapeutic targets, avoiding the possible complications ensuing from altered local production of steroid hormones. For example, overexpression of the endosomal cholesterol trafficking protein, STARD3, in human (THP-1) monocyte-macrophages, can also induce ABCA1 mRNA and protein, enhance cholesterol to ApoA-I, and limit cholesterol esterification in response to acetylated LDL, blocking cholesteryl ester deposition [56] (Fig. 5.1). However, it is unclear whether overexpression of STARD3 can also facilitate anti-inflammatory responses.

Equally, the nature of the stimulus or stimuli inducing the activity of expression of StAR in vascular tissues seems worthy of investigation. One possibility is that

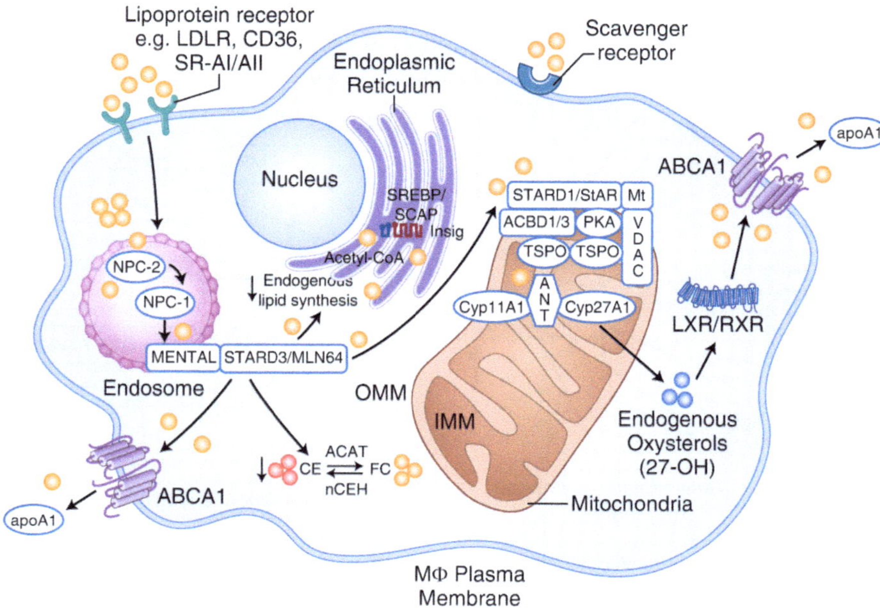

Fig. 5.1 Role of STARD1 (StAR) and STARD3 (MLN64) in maintaining macrophage cholesterol metabolism. Diagram illustrating the potential roles of STARD1 (StAR) and STARD3 (MLN64) in maintaining macrophage (MΦ) cholesterol metabolism. The over-expression of STARD1 drives mitochondrial cholesterol trafficking to sterol 27-hydroxylase (CYP27A1), enhancing the endogenous production of oxysterols (27-hydroxycholesterol, 27-OH), in turn activating the Liver X Receptor (LXR) and driving cholesterol efflux to apolipoprotein-A1 (apoA1), via ATP-binding cassette transporter A1 (ABCA1). The current model for cholesterol transfer from the outer (OMM) to inner (IMM) mitochondrial membrane involves a complex consisting of 18 kDa translocator protein (TSPO), VDAC (voltage dependent anion channel) and adenine nucleotide transporter (ANT). Over-expression of STARD3 may drive the transport of cholesterol to the plasma membrane, enhancing cholesterol efflux, via ABCA1, to lipid poor acceptors, such as ApoA1, or alternatively deliver endosomal cholesterol to the endoplasmic reticulum, retaining the Sterol regulatory element binding protein (SREBP)/SREBP-cleavage activating protein (SCAP) complex, in turn reducing cholesterol biosynthesis. Additionally, over-expression of STARD3 prevents increases in cholesterol esterification, reducing cholesteryl ester (CE) deposition *PKA* protein kinase A, *Acyl CoA* cholesterol ACAT, *ACAT* acyltransferase, *nCEH* neutral cholesteryl ester hydrolase, *FC* free cholesterol, *NPC-1/-2* Niemann-Pick C1/C2 protein, *StAR* steroidogenic acute regulatory protein

ApoA-I may contribute to this process, as ApoA-I can activate cellular cAMP signalling via the ABCA1 transporter, triggering serine phosphorylation of ABCA1; notably, this effect of ApoA-I is particularly evident in cells expressing high levels of ABCA1 [78]. Many factors regulate the expression, activity and phosphorylation status of StAR in steroidogenic tissues, including gonadotropins, cytokines, growth factors and environmental factors [48, 51, 61] (Fig. 5.2). Whether these factors impact on vascular StAR, the cholesterol efflux pathway or atherogenesis, in the same way is currently unknown. Intriguingly, gonadotropins, such as lutein-

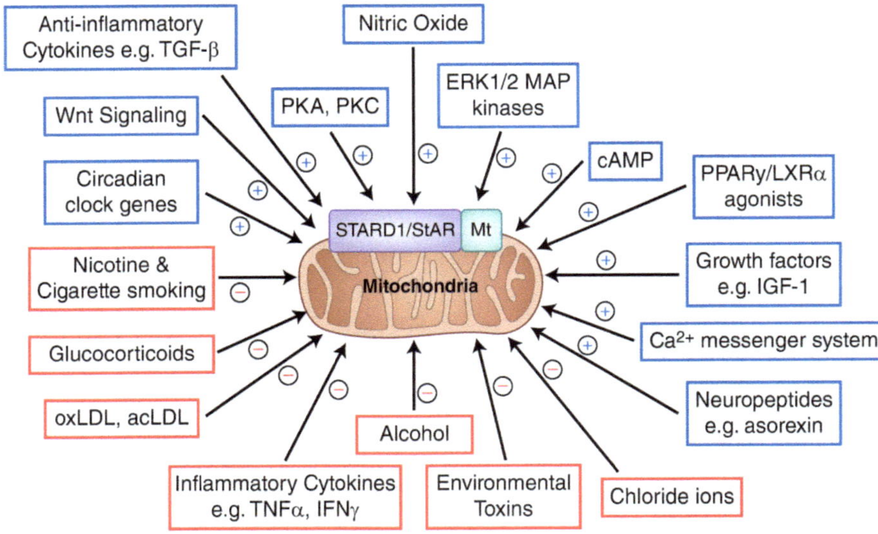

Fig. 5.2 Factors that influence the expression of steroidogenic acute regulatory (StAR) protein. Diagram illustrating intracellular and external factors that have the potential to regulate the expression and activity of StAR (STARD1) protein

izing hormone (LH) and/or follicle stimulating hormone (FSH), key regulators of StAR expression in steroidogenic tissues, increase macrophage expression of genes/proteins implicated in cholesterol efflux (LXRα, ABCA1, ABCG1, ABCG4, apolipoprotein (Apo)E) and stimulate cholesterol efflux to ApoA-I and high-density lipoprotein (HDL)[2] findings which may have resonance for cardiovascular health in individuals with altered endocrine function. Further, the cholesterol trafficking activity of StAR, and the output of pregnenolone in steroidogenic tissues is modulated by mitochondrial (dys)function [reviewed in 79], a feature associated with many atherosclerotic lesions [80, 81].

Finally, it seems sensible to acknowledge any deleterious effects which may result from increased expression or activity of StAR in vascular tissues. One obvious possibility is that the activation of LXRα will increase expression of SREBP-1c, resulting in induction of lipogenesis [27]. Indeed, Taylor et al. [65] reported StAR overexpression increased incorporation of [^{14}C]acetate into phospholipid (~2-fold), cholesterol (~2.5-fold) and triglyceride (~4-fold) pools, despite the fact that macrophage triglyceride mass did not change [65]. Further, 27-hydroxycholesterol can act as a selective estrogen receptor modulator (SERM), which binds to and modifies estrogen receptor (ER)α and β function, and may adversely impact on estrogen-related cardiovascular protection and/or bone mineralisation[81–84]. In rats, at least, 27-hydroxycholesterol inhibited the estrogen-dependent production of nitric oxide, reducing estrogen-mediated vasorelaxation [83].

[2] Borthwick Faye, Graham Annette (2009) Gonadotropin-dependent regulation of genes involved in the cholesterol efflux pathway. Glasgow Caledonian University.

In conclusion, accumulating evidence suggests alternative roles for StAR in non-steroidogenic tissues [51], including those within the vasculature [52–58; 62–65; 70, 74], but it is evident that studies in vivo are needed to fully evaluate the function of StAR in this context. To date, overexpression studies in vascular cells suggest an anti-atherogenic role for StAR, which could therefore represent a novel therapeutic target for treatment of one of the most prevalent causes of morbidity and mortality worldwide.

References

1. Libby P, Ridker PM, Hansson GK. Progress and challenges in translating the biology of atherosclerosis. Nature. 2011;473:317–25.
2. Woollard KJ. Immunological aspects of atherosclerosis. Clin Sci. 2013;125:221–35.
3. Wolfs IMJ, Donners MMPC, de Winther MPJ. Differentiation factors and cytokines in the atherosclerotic plaque micro-environment as a trigger for macrophage polarisation. Thromb Haemost. 2011;106:763–71.
4. Gerrity RG. The role of the monocyte in atherogenesis. Am J Pathol. 1981;103:181–90.
5. Sawamura T, Kakino A, Fujita Y. LOX-1: a multiligand receptor at the crossroads of response to danger signals. Curr Opin Lipidol. 2012;23:439–45.
6. Kzhyshkowska J, Neyen C, Gordon S. The role of macrophage scavenger receptors in atherosclerosis. Immunobiology. 2012;217:492–502.
7. Andres V, Pello O, Silvestre-Roig C. Macrophage proliferation and apoptosis in atherosclerosis. Curr Opin Lipidol. 2012;23:429–38.
8. Rye K-A, Barter PJ. Predictive value of different HDL particles for the protection against or risk of coronary heart disease. Biochim Biophys Acta. 2012;1821:473–80.
9. Tall AR. Cholesterol efflux pathways and other potential mechanisms involved in the athero-protective effect of high density lipoproteins. J Intern Med. 2008;263:256–73.
10. Gordon DJ, Knoke J, Probstfield JL, Superko R, Tyroler HA. High-density lipoprotein cholesterol and coronary heart disease in hypercholesterolaemic men: The Lipid Research Clinics Coronary Primary Prevention Trial. Circulation. 1986;74:1217–25.
11. Barter PJ, Nicholls S, Rye KA, Anantharamaiah GM, Navab M, Fogelman AM. Antiinflammatory properties of HDL. Circ Res. 2004;95:764−72
12. Scanu AM, Edelstein C. HDL: bridging past and present with a look to the future. FASEB J. 2008;22:4044−54
13. Soehnlen O, Swirski FP. Hypercholesterolaemia links haematopoiesis with atherosclerosis. Trends EndocrinolMetab. 2012. http://dx.doi.org.10.10106/j.tem.2012.10.008
14. Murphy AJ, Westerterp M, Yvan-Charvet L, Tall AR. Anti-atherogenic mechanisms of high density lipoprotein: effects on myeloid cells. Biochim Biophys Acta. 2012;1821:513–21.
15. Rosenson RS, Brewer HB Jr, Davidson WS, Fayad ZA, Fuster V, Goldstein J, Hellerstein M, Jiang XC, Phillips MC, Rader DJ, Remaley AT, Rothblat GH, Tall AR, Yvan-Charvet L. Cholesterol efflux and atheroprotection: advancing the concept of reverse cholesterol transport. Circulation. 2012;125:1905–19.
16. Cavalier C, Lorenzi I, Rohrer L, von Eckardstein A. Lipid efflux by the ATP binding cassette transporters, ABCA1 and ABCG1. Biochim Biophys Acta. 2006;1761:655–66.
17. Janowski BA, Willy PJ, Devi TR, Falck JR, Mangelsdorf DJ. An oxysterol signalling pathway mediated by the nuclear receptor LXRα. Nature. 1996;383:728–31.
18. Janowski BA, Grogan MJ, Jones SA, Wisely GB, Kliewer SA, Corey EJ, Mangelsdorf D. Structural requirements of ligands for the oxysterol liver X receptors LXRα and LXRβ. Proc Natl Acad Sci USA. 1999;96:266–71.

19. Venkateswaran A, Laffitte B, Joseph SB, Mak PA, Wilpitz DC, Edwards PA, Tontonez P. Control of cellular cholesterol efflux by the nuclear oxysterol receptor LXRα. Proc Natl Acad Sci USA. 2000;97:12097–102.
20. Wang Y, Rogers PM, Su C, Varga G, Stayrook KR, Burris YP. Regulation of cholesterologenesis by the oxysterol receptor LXRα. J Biol Chem. 2008;283:26332–9.
21. Zelcer N, Hong C, Boydjian R, Tontonez P. LXR regulates cholesterol uptake through Idol-dependent ubiquitination of the LDL receptor. Science. 2009;325:100–104.
22. Shibata N, Glass CK. Macrophages, oxysterols and atherosclerosis. Circ. J. 2010;74:2045–51.
23. Javitt NB. Oxysterols: novel biologic roles for the 21st century. Steroids. 2008;73:149–57.
24. Ghisletti S, Huang W, Ogawa S, Pascual G, Lin ME, Willson TM, Rosenfeld MG, Glass CK. Parallel SUMOylation-dependent pathways mediate gene- and signal-specific transrepression by LXRs and PPARγ. Mol Cell. 2007;25:57–70.
25. Im S-S, Osborne TF. Liver X receptors in atherosclerosis and inflammation. Circ Res. 2011;108:996–1001.
26. Joseph SB, Castrillo A, Laffitte BA, Mangelsdorf D, Tontonoz P. Reciprocal regulation of inflammation and lipid metabolism by liver X receptors. Nat Med. 2003;9:213–9.
27. Schultz JR, Tu H, Luk A, Repa JJ, Medina JC, Li L, Schwendner S, Wang S, Thoolen M, Mangelsdorf DJ, Lustig KD, Shan B. Role of LXRs in control of lipogenesis. Genes Dev. 2000;14:2831–8.
28. Cha JY, Repa JJ. The liver X receptor (LXR) and hepatic lipogenesis. The carbohydrate-response element binding protein is a target gene of LXR. J Biol Chem. 2007;282:743–51.
29. Bjorkhem I, Diczfalusy U, Lutjohann D. Removal of cholesterol from extrahepatic sources by oxidative mechanisms. CurrOpin. Lipidol. 1999;10:161–5.
30. Fu X, Menke G, Chen Y, Zhou G, Macnaul KL, Wright SD, Sparrow CP, Lund EG. 27-hydroxycholesterol is an endogenous ligand for liver X receptor in cholesterol-loaded cells. J Biol Chem. 2001;276:38373–87.
31. Lund E, Bjorkhem I, Furster C, Wikvall K. 24-, 25- and 27-hydroxylation of cholesterol by a purified preparation of 27-hydroxylase form pig liver. Biochim Biophys Acta. 1993;1155:177–82.
32. Chen W, Chen G, Head DL, Mangelsdorf DJ, Russell DW. Enzymatic reduction of oxysterols impairs LXR signalling in cultured cells and the livers of mice. Cell Metab. 2007;5:73–9.
33. Szanto A, Benko S, Szatmari I, Balint BL, Furtos I, Ruhl R, Molnar S, Csiba L, Garuti R, Calandra S, Larsson H, Diczfalusy U, Nagy I. Transcriptional regulation of human CYP27 integrates retinoid, peroxisome proliferator-activated receptor, and liver X receptor signalling in macrophages. Mol Cell Biol. 2004;24:8154–66.
34. Blorn T, Back N, Mutka AL, Bittman R, Li Z, de Lera A, Kovanen PT, Diczfalusy U, Ikonen E. FTY720 stimulates 27-hydroxycholesterol production and confers atheroprotective effects in human primary macrophages. Circ Res. 2010;106:720–9.
35. Zhu Y, Liao H, Xie X, Yuan Y, Lee TS, Wang N, Wang X, Shyy JY, Stemerman MB. Oxidized LDL downregulates ATP binding cassette transporter-1 in human vascular endothelial cells via inhibiting liver X receptor. Cardiovasc Res. 2005;68:425–32.
36. Brauner R, Johannes C, Ploessl F, Bracher F, Lorenz RI. Phytosterols reduce cholesterol absorption by inhibition of 27-hydroxycholesterol generation, LXRα activation and expression of the basolateral sterol exporter ATP binding cassette transporter A1 in Caco-2 enterocytes. J Nutr. 2012;142:981–9.
37. Li T, Chen W, Chiang JY. PXR induces CYP27A1 and regulates cholesterol metabolism in the intestine. J Lipid Res. 2007;48:373–84.
38. Kim WS, Chan SL, Hill AF, Guillemin GJ, Garner B. Impact of 27-hydroxycholesterol on amyloid beta peptide production and ATP-binding cassette transporter expression in primary human neurons. J Alzheimers Dis. 2009;16:121–31.
39. Escher G, Krozowski Z, Croft KD, Sviridov D. Expression of sterol 27-hydroxylase (CYP27A1) enhances cholesterol efflux. J BiolChem. 2003;278:11015–9.

40. Zhang JR, Coleman T, Langmade SJ, Scherrer DE, Lane L, Lanier MH, Feng C, Sands MS, Schaffer JE, Semenkovich CF, Ory DS. Niemann Pick C1 protects against atherosclerosis in mice via regulation of macrophage intracellular cholesterol trafficking. J. Clin. Invest. 2008;118:2281–90.
41. Bowden KL, Bilbey NJ, Bilawchuk LM, Boadu E, Sidhu R, Ory DS, Du H, Chan T, Francis GA. Lysosomal acid lipase deficiency impairs regulation of ABCA1 gene and formation of high density lipoproteins in cholesteryl ester storage disease. J Biol Chem. 2011;286:30624–35.
42. Gallus GN, Dotti MT, Federico A. Clinical and molecular diagnosis of cerebrotendinous xanthomatosis with a review of the mutations in the CYP27A1 gene. Neurol Sci. 2006;27:143–9.
43. Valdivielso P, Calandra S, Duran JC, Garuti R, Herrera E. Coronary heart disease in a patient with cerebrotendinus xanthomatosis. J Int Med. 2004;255:680–3.
44. Lorbek G, Lewinsha M, Rozman D. Cytochrome P450 s in the synthesis of cholesterol and bile acids - from mouse models to human diseases. FEBS J. 2012;279:1516–33.
45. Pandak WM, Ren S, Marques D, Hall E, Redford K, Mallonee D, Bohdan P, Heuman D, Gil G, Hylemon P. Transport of cholesterol into mitochondria is rate-limiting for bile acid synthesis via the alternative pathway in primary rat hepatocytes. J Biol Chem. 2002;277:48158–64.
46. Bjorkhem I, Diczfalusy U, Lutjohann D. Removal of cholesterol from extrahepatic sources by oxidative mechanisms. Curr Opin Lipidol. 1999;10:161–5.
47. Rone MB, Fan J, Papadopoulos V. Cholesterol transport in steroid biosynthesis: role of protein-protein interactions and implications in disease states. Biochim Biophys Acta. 2009;179:646–58.
48. Manna PR, Dyson MT, Stocco DM. Regulation of the steroidogenic acute regulatory protein gene expression: present and future perspectives. Mol Hum Reprod. 2009;6:321–33.
49. Miller WL. Bose HS. Early steps in steroidogenesis: intracellular cholesterol trafficking. J Lipid Res. 2011;52:2111–35.
50. Rone MB, Midzak AS, Issop L, Rammouz G, Jagannathan S, Fan J, Ye X, Blonder J, Veenstra T, Papadopoulos V. Identification of a dynamic mitochondrial protein complex driving cholesterol import, trafficking, and metabolism to steroid hormones. MolEndocrinol. 2012;26:1868–82.
51. Anuka E, Gal M, Stocco DM, Orly J. Expression and roles of steroidogenic acute regulatory (StAR) protein in 'non-classical', extra-adrenal and extra-gonadal cells and tissues. Mol Cell Endocrinol. 2013;371:47–61.
52. Ning Y, Chen S, Li X, Ma Y, Zhao F, Yin L. Cholesterol, LDL and 25-hydroxycholesterol regulate expression of the steroidogenic acute regulatory protein in microvascular endothelial cell line (bEnd.3). Biochem. Biophys Res Commun. 2006;342:1249–56.
53. Castillo SS, Doger MM, Bolkent S, Yanardag R. Cholesterol efflux and the effect of combined treatment with niacin and chromium on aorta of hyperlipidaemic rat. Mol Cell Biochem. 2007. doi:10.1007/s11010-007-9623-2.
54. Ma Y, Ren S, Pandak WM, Li X, Ning Y, Lu C, Zhao F, Yin L. The effects of inflammatory cytokines on steroidogenic acute regulatory protein expression in macrophages. Inflamm Res. 2007;56:495–501.
55. Borthwick F, Taylor JM, Bartholomew C, Graham A. Differential regulation of the STARD1 subfamily of START lipid trafficking proteins in human macrophages. FEBS Lett. 2009;583:1147–53.
56. Borthwick F, Allen A-M, Taylor JM, Graham A. Overexpression of STARD3 in human monocyte/macrophages induces an anti-atherogenic lipid phenotype. Clin Sci. 2010;119:265–72.
57. Christenson LK, Osborne TF, McAllister JM, Strauss JF III. Conditional response of the human steroidogenic acute regulatory protein gene promoter to sterol regulatory element binding protein-1a. Endocrinology. 2001;142:28–36.
58. Castillo SS, Doger MM, Bolkent S, Yanardag R. Cholesterol efflux and the effect of combined treatment with niacin and chromium on aorta of hyperlipidaemic rat. Mol Cell Biochem. 2008;308:151–9.

59. Niu N, Yu H, Wang Y, Wang LJ, Li Q, Guo M. Combined effects of niacin and chromium treatment on vascular endothelial dysfunction in hyperlipidaemic rats. Mol Biol Rep. 2009;36:1275–81
60. Kamanna VS, Kashyap KL. Mechanism of action of niacin. Am J Cardiol. 2008;101(supple):20B–26B.
61. Manna PR, Dyson MT, Stocco DM. Regulation of the steroidogenic acute regulatory protein gene expression: present and future perspectives. Mol Hum Reprod. 2009;15:321–33.
62. Ning Y, Ren S, Zhao F, Yin L. Overexpression of the steroidogenic acute regulatory protein increases the expression of ATP-binding cassette transporters in microvascular endothelial cells (bEnd.3). J Zhejiang Univ Sci B. 2010;11:350–6.
63. Tian D, Qiu Y, Zhan Y, Li X, Wang X, Yin L, Ning Y. Overexpression of steroidogenic acute regulatory protein in rat aortic endothelial cells attenuates palmitic acid-induced inflammation and reduction in nitric oxide bioavailability. Cardiovasc Diabetol. 2012;11:144. doi:10.1186/1475-2849-11-144
64. Ning Y, Bai Q, Lu H, Li X, Pandak WM, Zhao F, Chen S, Ren S, Yin L. Overexpression of mitochondrial cholesterol delivery protein, StAR, decrease intracellular lipids and inflammatory factors secretion in macrophages. Atherosclerosis. 2009;204:114–20.
65. Taylor JMW, Borthwick F, Bartholomew C, Graham A. Overexpression of steroidogenic acute regulatory protein increases macrophage cholesterol efflux to apolipoprotein AI. Cardiovasc Res. 2010;86:526–34.
66. Mack JT, Brown CB, Garrett TE, Uys JD, Townsend DM, Tew KD. Ablation of the ATP-binding cassette transporter, Abca2, modifies response to estrogen-based therapies. Biomed and Pharmacother. 2012;66:403–8.
67. Calpe-Berdiel, L, Zhao Y, de Graauw M, Ye D, van Santbrink PJ, Mommaas AM, Foks A, Bot M, Meurs I, Kuiper J, Mack JT, Van Eck M, Tew KD, van Berkel TJC. Macrophage ABCA2 deletion modulates intracellular cholesterol deposition, affects macrophage apoptosis and decreases early atherosclerosis in LDL receptor knockout mice. Atherosclerosis. 2012;223:332–41.
68. Ogita M, Miyauchi K, Dohi T, Tsuboi S, Miyazaki T, Yokoyama T, Yokoyama K, Shimada K, Kurata T, Jiang M, Bujo H, Daida H. Increased circulating soluble LR11 in patients with acute coronary syndrome. ClinChimActa. 2013;415:191–4.
69. Sheikine Y, Sirsjo A. CXCL16/SR-PSOX—a friend or foe in atherosclerosis? Atherosclerosis. 2008;197:487–95.
70. Bai Q, Li X, Ning Y, Zhao F, Yin L. Mitochondrial cholesterol transporter, StAR, inhibits human THP-1 monocyte-derived macrophage apoptosis. Lipids. 2009;45:29–36.
71. Hasegawa T, Zhao L, Caron KM, Majdic G, Suzuki T, Shizawa S, Sasano H, Parker KL. Developmental roles of the steroidogenic acute regulatory protein (StAR) as revealed by StAR knockout mice. Mol Endocrinol. 2000;14:1462–71.
72. Ishii T, Hasegawa T, Pai CI, Yvgi-Ohana N, Timberg R, Zhao L, Majdic G, Chung BC, Orly J, Parker KI. The roles of circulating high-density lipoproteins and trophic hormones in the phenotype of knockout mice lacking the steroidogenic acute regulatory protein. Mol Endocrinol. 2002;16:2297–309.
73. Ishii T, Mitsui T, Suzuki S, Matsuzaki Y, Hasegawa T. A genome-wide expression profile of adrenocortical cells in knockout mice lacking steroidogenic acute regulatory protein. Endocrinology. 2012;153:2714–23.
74. Ning Y, Xu L, Ren S, Pandak WM, Chen S, Yin L. StAR overexpression decreases serum and tissue lipids in apolipoprotein E-deficient mice. Lipids. 2009;44:511–9.
75. Young MJ, Clyne CD, Cole TJ, Funder JW. Cardiac steroidogenesis in the normal and failing heart. J Clin Endocrinol Metab. 2001;86:5121–6.
76. Casal AJ, Silvestre JS, Delcayre C, Capponi AM. Expression and modulation of steroidogenic acute regulatory protein messenger ribonucleic acid in rat cardiomyocytes and after myocardial infarction. Endocrinology. 2003;144:1861–8.

77. Anuka A, Yiygi-Ohana N, Eimerl S, Garfinkel B, Melamed-Book N, Chepurkol E, Aravot D, Zinman T, Shainberg A, Hochhauser E, Orly J. Infarct-induced steroidogenic acute regulatory protein: a survival role in cardiac fibroblasts. Mol Endocrinol. 2013;27:1502–17.
78. Haidar B, Denis M, Marcil M, Krimbou L, Genest J Jr. Apolipoprotein A-I activates cellular cAMP signalling through the ABCA1 transporter. J Biol Chem. 2004;279:9963–9.
79. Allen AM, Taylor JMW, Graham A. Mitochondrial (dys)function and regulation of macrophage cholesterol efflux. Clin Sci. 2013;124:509–15.
80. Sobenin IA, Chistiakov DA, Bobryshev YB, Postnov AY, Orekhov AN. Mitochondrial mutations in atherosclerosis: new solutions in research and possible clinical applications. Curr Pharm Des. 2013 Feb 15. Epub ahead of print.
81. Madamanchi NR, Runge MS. Mitochondrial dysfunction in atherosclerosis. Circ Res. 2007;100:460–73.
82. Umetani M, Domoto H, Gormley AK, Yuhanna IS, Cummins CL, Javitt NB, Korach KS, Shaul PW, Mangelsdorf DJ. 27-hydroxycholesterol is an endogenous SERM that inhibits the cardiovascular effects of estrogen. Nat Med. 2007;13:1185–92.
83. Umetani M, Shaul PW. 27-hydroxycholesterol: the first identified endogenous SERM. Trends Endocrinol Metab. 2011;22:130–5.
84. Baker ME. What are the physiological estrogens? Steroids. 2013;78:337–40.

Chapter 6
STARD3: A Lipid Transfer Protein in Breast Cancer and Cholesterol Trafficking

Fabien Alpy and Catherine L. Tomasetto

Abstract STARD3 was isolated in the early 1990s in a study aimed at finding new genes implicated in breast cancer. The function of the STARD3 gene, referred to at that time as Metastatic Lymph Node clone number 64 (MLN64), remained a mystery until the discovery of the steroidogenic acute regulatory protein (StAR/STARD1). Indeed, homology searches showed a region of significant similarity between StAR and the carboxy-terminal half of STARD3. This homology proved to be functionally relevant with both proteins being cholesterol carriers; however, quite early it appeared that they were very distinct in terms of expression, subcellular localization, and function. It was then reported that STARD3 was part of a family of 15 human proteins that shared a conserved StAR-related lipid transfer (START) domain. Structurally, the STARD3 protein distinguishes itself by the presence of an additional conserved domain spanning the amino-terminal half of the protein that we named the MLN64-N-terminal (MENTAL) domain. This domain contains most of the functional properties that have been attributed to STARD3. This chapter will present our current understanding of STARD3 function in cancer, cell biology, and cholesterol trafficking.

Abbreviations

CAB1:	Co-Amplified with ERBB-2
CBP:	Carotenoid binding protein
CHO:	Chinese Hamster ovary
DARPP32:	Dopamine and cAMP-regulated phosphoprotein
EGFR:	Epidermal growth factor receptor
ER:	Endoplasmic reticulum
FFAT:	Diphenylalanine [FF] in an acidic tract

C. L. Tomasetto, Ph.D. (✉) · F. Alpy, Ph.D.
Department of Functional Genomics and Cancer, Institut de Génétique et de Biologie Moléculaire et Cellulaire (IGBMC), Institut National de la Santé et de la Recherche Médicale (INSERM), U 964—Centre National de la Recherche Scientifique (CNRS) UMR 7104—University of Strasbourg, 1 rue Laurent Fries, 67404 Illkirch, France
e-mail: Catherine-Laure.Tomasetto@igbmc.fr

F. Alpy
e-mail: Fabien.Alpy@igbmc.fr

B. J. Clark, D. M. Stocco (eds.), *Cholesterol Transporters of the START Domain Protein Family in Health and Disease,* DOI 10.1007/978-1-4939-1112-7_6,

GRB7:	Growth factor bound protein 7
HER2:	Epidermal growth factor receptor 2
LE:	Late endosome
LOF:	Loss-of-function
LY:	Lysosome
MCS:	Membrane contact site
MENTAL:	MLN64-N-terminal domain
MENTHO:	MLN64 N-terminal domain homologue
MLN64:	Metastatic lymph node clone 64
NPC1:	Niemann-Pick C1
SRA:	Smallest region of amplification
StAR:	Steroidogenic acute regulatory protein
STARD3:	START domain containing protein 3
STARD3NL:	STARD3 N-terminal like
START:	(StAR)-related lipid transfer
TEM:	Transmission electron microscopy
VAP:	VAMP (vesicle-associated membrane protein)-associated protein

Introduction

STARD3 Perspectives: How we Discovered the Second Member of the Steroidogenic Acute Regulatory Protein (StAR)-Related Lipid Transfer (START) Protein Superfamily by Catherine L. Tomasetto

This is my personal recollection about the discovery of STARD3 more than 20 years ago. I remember it as a collective effort from several people including two young brilliant graduate students Dr. Catherine Regnier and Dr. Christel Moog. It all started in the early 1990s when I came back to Strasbourg after spending 3 years in Boston at Harvard University, in the Dana-Farber Cancer Institute. Under Dr. Ruth Sager's firm guidance, my colleagues and I chased tumor suppressor genes in mammary epithelial cells. There I learned new methods in cloning and manipulating genes; importantly, I became an expert in the subtractive hybridization method. In Strasbourg, I joined one of the best research institutes in Europe—the Institut de Génétique et de Biologie Moléculaire et Cellulaire (IGBMC)—a pioneer in many aspects of molecular biology especially in the field of gene expression. The IGBMC was led by Dr. Pierre Chambon, one of the most prominent life scientists. In this large institute, I naturally joined the group working on breast cancer headed by Dr. Paul Basset and Dr. Marie-Christine Rio. My project was to find new genes important for breast cancer, which could be used as prognostic markers, therapeutic targets, and which would help to understand the biology of cancer. The idea was to isolate genes expressed in breast tumors but not in normal breast tissue. This was before the microarray era and only tedious methods were available. I chose subtractive hybridization and we decided to work on patient biopsies instead of established cell lines. Because we were interested in finding genes involved in aggressive tumors, we selected tumor samples using the following clinical criteria: young age of the patient at the onset of diagnosis, large tumor size, and elevated histological grade. After 2 years of screening, cloning, and sequencing, we narrowed down the number of positives hits from over 200 to 10 independents genes [1]. Among these 10 genes, the clone number 64 was intriguing because its sequence was novel, it was overexpressed in all breast tumors that were positive for the epidermal growth factor receptor 2 (HER2) oncogene and it was expressed at a basal level in all tissues and cell lines tested. Christel Moog undertook the challenge of characterizing this novel gene; she notably found the homology with steroidogenic acute regulatory protein (StAR) and produced the consensus sequence for this region of homology that she called the StAR homology domain (SHD; [2]). We tried to establish a link between STARD3 and endocrine-dependent breast cancer. One of the major subtypes of breast cancer expresses high levels of the estrogen receptor and can be treated with endocrine therapies. To our disappointment, Christel found that STARD3 did not promote steroidogenesis in breast cancer cells and was not localized in the mitochondria (unpublished, [2]). Then, fortunately, a dedicated and talented graduate student, Dr. Fabien Alpy, joined us, became interested in this peculiar protein, and took on the challenge of finding its function. His major contributions were to discover the subcellular localization of STARD3 in late endosomes and the definition of a novel conserved domain that he called MENTAL for MLN64-N-terminal domain [3]. Almost a decade later, Fabien is now a staff scientist on the team, remains very interested in STARD3 and is committed to the pathophysiological functional characterization of STARD3.

I thank all the colleagues who have contributed to the work that is described in this chapter. I apologize for leaving out many names. I acknowledge funding from the French research agencies: Institut National de la Santé et de la Recherche Médicale (INSERM), Center National de la Recherche Scientifique (CNRS); the University of Strasbourg; and the charities: the Fondation pour la Recherche Medicale, the fondation ARC pour la recherche sur le cancer, and the Ligue nationale Contre le Cancer.

STARD3 Overview

STARD3 was first isolated in a screen designed to identify new genes implicated in breast cancer. Using subtractive hybridization as a screening method, several unknown complementary deoxyribonucleic acids (cDNAs) were isolated from a library prepared from a pool of metastatic lymph nodes derived from breast cancers. STARD3 was originally the 64th clone isolated in the screen and was therefore named metastatic lymph node clone 64 (MLN64; [1]). In this original publication, we showed that the STARD3 gene lied next to the oncogene HER2/ERBB2 on chromosome 17q11-12 and proposed that amplification of this chromosomal subregion was responsible for the observed co-expression between STARD3 and HER2 in breast cancer cells [1]. This hypothesis was substantiated using a series of about 100 human primary breast cancers which showed that STARD3 was indeed co-amplified and co-expressed with HER2 in about 25 % of breast cancers [4]. Several years later, STARD3 was isolated again by a different group in the context of breast cancer and named CAB1 for co-amplified with ERBB-2 as a new gene co-amplified and overexpressed in breast cancer [5]. Soon after its identification, we noted that STARD3 shares a functionally conserved domain with the StAR protein [2, 6]; this finding provided the foundation for the definition of a novel conserved protein family called the StAR-related lipid transfer (START) protein superfamily ([7–11] and see Chaps. 1 and 2). A second insight into STARD3's function came from the identification of its subcellular localization at the membrane of late endosomes [12] and from its bi-functional protein structure [13, 14]. Over the last decade, efforts to understand the pathophysiological function of STARD3 have proceeded along two main research themes, cholesterol trafficking and cancer; these themes will be discussed in this chapter.

STARD3 is a Bi-Functional Protein Conserved from Unicellular Organisms to Humans

STARD3 is characterized by the presence of two distinct structural domains separating the protein in two halves (Fig. 6.1a). The amino-terminal half of the protein is highly conserved during evolution; it is present in a second protein called STARD3 N-terminal like (STARD3NL) originally named MLN64 N-terminal homolog (MENTHO; [13]). As shown in Fig. 6.1a, the carboxy-terminal half contains a START domain conserved with STARD1 also known as StAR (see Chap. 1). Like STARD1, STARD3 can enhance steroidogenesis in cellular assays; removal of the conserved START domain resulted in the complete loss of steroidogenic activity; on the contrary, removal of the amino-terminal region of STARD3 increased this activity [6, 15]. In addition, the isolated recombinant START domain of STARD3 was shown to be a sterol transfer protein in vitro [16–18]. In early studies, the 3-dimensional structure of the START domain of STARD3 illustrated the presence of a deep lipid-binding pocket able to accommodate one molecule of cholesterol; consistently, in vitro the START domain bound cholesterol at an equimolar ratio [19]. Since the structure/function of the START domain will be extensively described in other chapters (see Chaps. 3 and 4), we will describe here mainly the structure/function of the MENTAL domain of STARD3.

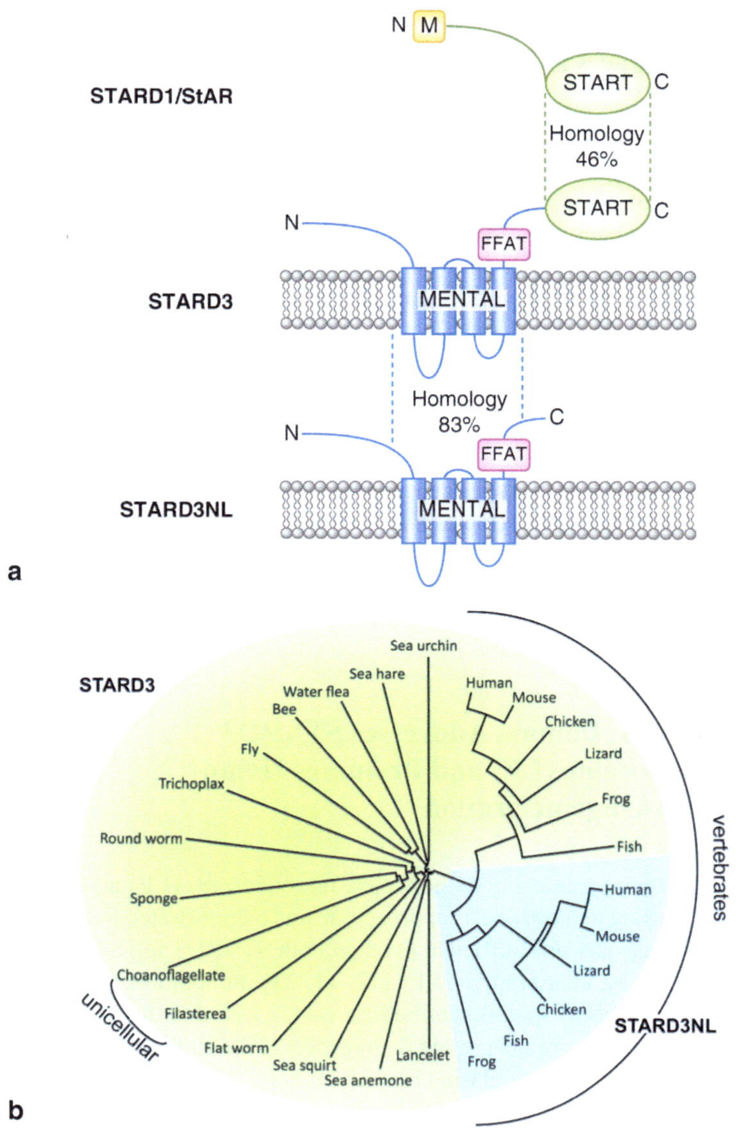

Fig. 6.1 STARD3 contains two domains and is conserved during evolution. **a** Schematic representation of STARD1, STARD3 and STARD3NL proteins. The START and MENTAL domains are in *green* and *blue* respectively. Transmembrane helices within the MENTAL domain are boxed in *blue*. The mitochondrial-addressing (M) and the FFAT motifs are boxed in *yellow* and *pink*, respectively. The amino- and carboxy-terminal extremities are indicated by N and C, respectively. Numbers located between *dotted lines* are similarity percentages between the proteins in the domain defined. **b** Phylogenetic tree of 20 STARD3 and 6 STARD3NL orthologs. The whole sequences were aligned by the Eclustalw program (Genetics Computer Group, Madison, WI). The phylogenetic tree was drawn with the Interactive Tree Of Life v2 software [67]. *START* steroidogenic acute regulatory protein (StAR)-related lipid transfer, *MENTAL* metastatic lymph node clone number 64-N-terminal, *FFAT* diphenylalanine [FF] in an acidic tract

The MENTAL Domain Distinguishes STARD3 from the Other START Proteins

The finding of STARD3NL, a second protein sharing a high homology with the N-terminal half of STARD3, provided the basis for the definition of a novel protein domain that we named MENTAL after the original designation of STARD3 as the MLN64 protein [13, 14]. The MENTAL domain is composed of four transmembrane helices with three short intervening loops (Fig. 6.1a). This organization resembles the structural organization of the protein from the Tetraspanin superfamily; however, no significant homology with Tetraspanin protein sequences or with other proteins containing four transmembrane helices was found. Thus, the MENTAL domain is unique to STARD3 and STARD3NL. To examine the evolution of STARD3, we performed a multi-alignment analysis of STARD3 and STARD3NL primary sequences (Fig. 6.1b). Interestingly, STARD3 is present within all the animal kingdom as well as in unicellular organisms closely related to animals [20, 21]. Unlike STARD3, STARD3NL exists only in vertebrates (Fig. 6.1b). The restricted presence of STARD3NL in vertebrates as well as the identical gene organization between human STARD3 and STARD3NL genes [22] suggest that STARD3NL originates from a duplication event of the STARD3 gene.

The MENTAL Domain Addresses STARD3 to Late Endosome (LE) and Promotes Homo- and Hetero-Oligomerization

STARD3 does not contain any of the typical late endosome (LE)-addressing motifs. However, by using a mutagenesis approach, it was demonstrated that the MENTAL domain was necessary and sufficient to target both STARD3 and STARD3NL to the endosomal limiting membrane of LEs [12, 13, 23]. In addition, using microinjection and endocytosis of specific antibodies, we showed that the MENTAL domain controls the orientation of the protein with respect to the endosomes and projects the START domain into the cytoplasm [12, 13]. Of note, utilizing electron microscopy, STARD3 was shown to mainly localize at the surface of late endosomes with an uneven staining pattern, suggesting that the protein could accumulate in discrete subregions of the membrane [12]. The structural resemblance of the MENTAL domain with Tetraspanins, which are known to associate with one another and to form a tetraspanin web [24] also supported the notion that STARD3 and STARD3NL could form protein complexes at the membranes of endosomes. Using classical biochemistry and imaging approaches, we demonstrated that STARD3 and STARD3NL homo- and hetero-oligomerize and that the MENTAL domain is instrumental in promoting this interaction [14].

The MENTAL Domain is a Cholesterol-Binding Domain

To gain insight into the molecular mechanism implicated in the handling of cholesterol by STARD3, an in vivo cholesterol-binding assay was used in living cells. In this experiment, radiolabeled photoactivatable cholesterol was provided to living cells using low-density lipoprotein particles (LDL). After cross-linking and immunoprecipitation, it was shown that the MENTAL domain of STARD3 was a cholesterol-binding domain indicating that STARD3 contains two distinct cholesterol-handling domains [14]. Moreover, a recent proteome-wide mapping of cholesterol-interacting proteins substantiates this finding. Using the click chemistry methodology, the repertoire of cholesterol-bound peptides from HeLa cells has been identified by Hulce et al. [25]. Of interest, several cholesterol-bound peptides belonging to STARD3 (residues 243–260 and 286–307) and STARD3NL (residues 23–37 and 185–196) were identified. These novel findings support the notion that both STARD3 and STARD3NL, by using their MENTAL domains, trigger the formation of microdomains enriched in cholesterol. The exact molecular mechanism behind cholesterol transport mediated by STARD3 and STARD3NL is still unclear however as shown in Fig. 6.2a, several mode of action models can be proposed. The exact molecular mechanism behind cholesterol transport mediated by STARD3 and STARD3NL is still unclear. However, as shown in Fig. 6.2a, several mode of action models can be proposed.

The MENTAL Domain has an Endoplasmic Reticulum (ER) Tethering Activity We recently showed that the MENTAL domain contains a conserved diphenylalanine [FF] in an acidic tract (FFAT)-like motif ([26]; Fig. 6.1a). FFAT motifs were identified as signals responsible for targeting cytosolic proteins to the surface of the ER by directly interacting with vesicle-associated membrane protein (VAMP)-associated protein] proteins (VAP; [27]). For instance, some oxysterol-binding protein-related proteins (ORPs) contain conserved FFAT motifs that target them to the ER surface. Because STARD3 and STARD3NL are not free cytosolic proteins but LE-anchored proteins, the interaction between STARD3, STARDNL and VAP proteins has a strong impact on the subcellular architecture. In cells expressing either STARD3 or STARD3NL, transmission electron microscopy pictures show that endosomes are covered by the ER (Fig. 6.2b). To gain more details on the localization of STARD3 with respect to subcellular architecture, immunogold labeling of STARD3 was performed on ultrathin sections using an antibody coupled to gold particles (Fig. 6.2b). Labeling of STARD3 was enriched at the outer periphery of LE in regions directly facing the ER (Fig. 6.2b; [26]) This endosome-ER tethering results in the formation of specific subcellular regions named membrane contact sites (MCSs) (Fig. 6.2b). The molecular identity of this LE-ER tethering complex is not exhaustive, but it contains STARD3 and/or STARD3NL and VAP proteins. In addition, extended endosome-ER contacts mediated by STARD3 alter the dynamics of the endosomal compartment by preventing vesicle-to-tubule transitions [26]. The functional significance of these

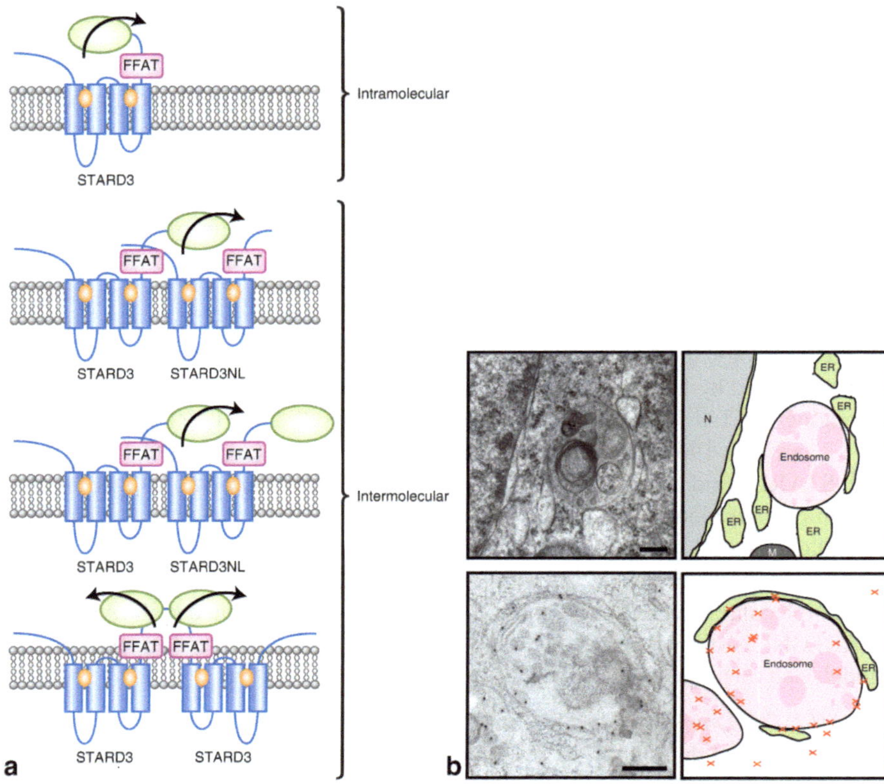

Fig. 6.2 STARD3 belongs to an inter-oganelle tethering machine. **a** Possible modes of action of STARD3 to transfer cholesterol. The MENTAL domain of STARD3 (*blue*) anchors the protein to endosome membranes, leaving its C-terminal START domain (*green*) in the cytoplasm. The START domain of STARD3 might work by extracting cholesterol (*yellow*) bound to its MENTAL domain from the late endosome membrane and transfering it to a closely positioned acceptor membrane or protein. STARD3 works as a monomer, oligomer, or inverted oligomer at the membrane of late endosomes making several cholesterol transfer scenarios possible. STARD3 may extract cholesterol using intramolecular or intermolecular mechanisms (see text for details). **b** STARD3 expression induces the wrapping of the ER around endosomes. *Top panels*: *Left*: transmission electron microscopy image of HeLa cells overexpressing STARD3. *Right*: interpretation scheme of the TEM image: the ER and endosomes are *green* and *pink*, respectively; part of the nucleus and part of a mitochondria are *gray* and *black*, respectively; ribosomes are represented as *small dots lining* the ER. *Bottom panels*: immunogold labeling of STARD3 positive endosomes. *Left*: *gold*-labeled STARD3 (*black dots*) is mainly found on the limiting membrane of endosomes. STARD3-positive endosomes are often tightly surrounded by ER-like structures. *Right*: Interpretation scheme of the TEM image with *gold* particles represented by *red crosses*; endosomes and ER are colored in *pink* and *green*, respectively. Bars represent 200 nm. *START* steroidogenic acute regulatory protein (StAR)-related lipid transfer, *MENTAL* metastatic lymph node clone number 64-N-terminal, *TEM* transmission electron microscopy, *ER* endoplasmic reticulum

specific endosome-ER MCSs remains to be addressed, but they probably represent specialized subcellular regions where signals and small molecules, such as lipids and calcium, are exchanged between organelles.

I. STARD3 is Implicated in the HER2 Molecular Subtype of Breast Cancer

STARD3 Belongs to the Smallest Region of HER2 Amplification Breast cancer is a heterogeneous disease at the genetic, molecular, and clinical levels. Studies reporting gene expression profiling revealed multiple tumor subtypes [28]. These distinct subtypes may reflect the variety of cell types found in the breast, their differentiation stages, and their phenotypic modifications via specific genomic alterations and key signaling pathways. In clinical practice, a basic molecular classification of breast cancer that considers three main subtypes is used to determine patient care: Luminal A and B, which express the estrogen receptor; HER2-positive tumors characterized by amplification and overexpression of the epidermal growth factor receptor 2 (ERBB2/HER2/neu); and triple-negative cancers, which do not express these receptors. This classification and the use of anti-hormonal and HER-2-targeted therapy made a significant difference in clinical outcome [29]. Nevertheless, within these groups heterogeneity remains and, for instance, the molecular variations within the HER2 subgroup and their clinical implications remain largely unknown. HER2-positive tumors have higher levels of overall genomic instability than HER2-negative tumors. This observation supports the notion that HER2 amplification is functionally implicated in chromosomal instability on Chr17q [30]. Moreover, the complexity of the genomic alterations found at the HER2 locus highlights the diversity of HER2-amplified breast cancers. Since the identification of HER2 amplification in breast cancer, multiple genes have been reported to be co-amplified with HER2, supporting the idea that they contribute to the phenotype of individual tumors (reviewed in [31]). By identifying four new genes overexpressed and co-amplified with HER2 in breast cancer, we made one of the first observations in this area and proposed that these co-amplified genes—in particular STARD3—could possibly contribute to HER2-positive cancers [1, 4].

When STARD3 was identified, it was clear that its expression in breast cancer was linked with HER2 amplification. We demonstrated that the co-amplification of STARD3 and HER2 is due to the close association of their genes [1], an association also conserved in the mouse genome (F.A. and C.T., unpublished). As shown in Fig. 6.3a, STARD3 and HER2 are located approximately 30–40 kb apart on chromosome 17q11-12. Several other genes are also found in this region that flank STARD3; on the centromeric side, the dopamine and cyclic adenosine monophosphate (cAMP)-regulated phosphoprotein (DARPP32) gene is 0.5 kb upstream of the STARD3 transcriptional start site, whereas on the telomeric side, the Telethonin gene starts 2 kb downstream of the 3′ end of the STARD3 gene [22]. However, unlike STARD3 and HER2, these two genes, although amplified, are not always expressed in HER2 breast cancer cells (Fig. 6.3b), indicating that amplification is not sufficient to drive overexpression [22]. Actually, the analysis of the STARD3 promoter region provided a clue regarding its overexpression with HER2. Indeed, both STARD3 and HER2 genes share Sp1 binding sites in their promoter regions supporting the notion that these genes are likely to be co-regulated by transcription

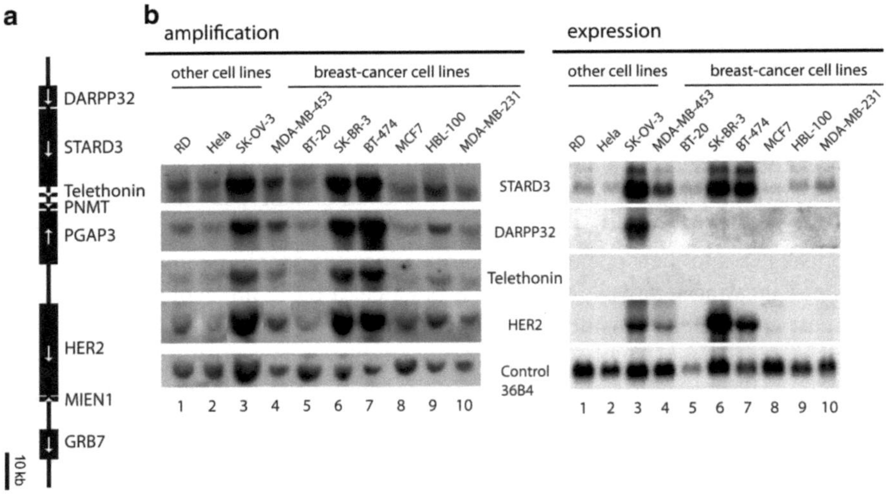

Fig. 6.3 Co-amplification of STARD3 with HER2 in breast cancer *cell lines*. (Adapted from [22]). **a** Schematic representation of the genes present in the HER2 amplicon. This map was built using the human chromosome 17 sequence. DARPP32, dopamine, and cAMP regulated phosphoprotein (Accession no. AF233349); STARD3 alias metastatic lymph node 64 (Accession no. NM 006804.3); TCAP, telethonin or titin cap (Accession no. NM 003673); PNMT, phenylethanolamine N-methyltransferase (Accession no. NM 002686); PGAP3, post-GPI attachment to proteins 3 (Accession no NM 033419); ERBB2, avian erythroblastic leukemia viral oncogene homolog 2 (Accession no. NM 004448); MIEN1, migration and invasion enhancer 1 (Accession no. NM 032339.3); GRB7, growth factor receptor-bound protein 7 (Accession no. XM 012695). **b** Amplification and expression analysis of STARD3, DARPP32, Telethonin, and HER2 in cancer *cell lines*. A total of 10 μg of EcoRI-digested genomic deoxyrhobonucleic acid (DNA) (i) or 10 μg of total ribonucleic acid (RNA) (ii) extracted from various cancer *cell lines* were loaded in each lane as indicated. Hybridizations were carried out successively with probes corresponding to STARD3, dopamine and cyclic adenosine monophosphate (cAMP)-regulated phosphoprotein (DARPP32), Telethonin, and epidermal growth factor receptor 2 (HER2). The loading control was the acidic ribosomal phosphoprotein P0, 36B4 probe

factors belonging to the specificity protein/Krüppel-like factor (Sp/KLF) family that bind to these motifs [22].

To date, several studies using genome-wide microarray methods have defined the molecular identity of the HER2 amplicon in breast cancer [32–37]. Taken together, these studies indicate that the HER2 smallest region of amplification (SRA) is limited to a small number of genes, including STARD3 and growth-factor-bound protein 7 (GRB7). The data suggest that co-amplification and co-expression of various genes of the HER2-SRA likely influence the biology of theses tumors, including the response to anti-cancer treatments. Therefore, it is necessary to understand the prognostic impact and function of STARD3 in HER2-positive cancers.

STARD3 Contributes to the HER2-Amplicon Addiction Phenotype Some studies have addressed the prognostic significance of STARD3 co-amplification in patients with HER2-positive tumors; they all agree in showing that STARD3 amplification is associated with a shorter overall survival and disease-free survival in these patients

[38–40]. The association of STARD3 with a poorer prognosis supported the idea that it has a direct influence on the aggressiveness of breast cancer cells. We originally proposed that STARD3 plays a synergistic role with HER2 during breast carcinogenesis. Consistent with this hypothesis, STARD3 downregulation in HER2 amplified cell lines resulted in reduced cell proliferation and increased cell death; on the contrary, STARD3 silencing in other cancer cells which are not HER2-amplified has no impact on cell growth and viability ([41]; Fig. 6.4a and b). Paradoxically, STARD3 forced expression in a non-HER2 cell line, like HeLa cells, compromises cell proliferation (Fig. 6.4a and b). The molecular mechanism by which STARD3 and HER2 may cooperate is still unclear. STARD3 and HER2 are highly conserved across different species and they are ubiquitously expressed but belong to distinct protein families. HER2 belongs to the epidermal growth factor receptor (EGFR) protein family that comprises four members, EGFR, HER2, HER3, and HER4 [42]. Unlike the other member of the family, HER2 lacks ligand-binding activity. Moreover, compared to EGFR which undergoes a robust endocytosis and degradation upon activation, HER2 is slowly endocytosed and recycled to the cell surface [43]. Despite many attempts, we failed to find a direct interaction between STARD3 and HER2 . Actually, in HER2-positive tumor samples, these proteins do not co-localize, with HER2 being predominantly present at the cell surface and STARD3 in the late endocytic compartment (Fig. 6.4c). Therefore, we reasoned that STARD3 might cooperate with HER2 by an indirect molecular mechanism which remains to be identified. Of interest, through loss of function studies (LOF), Cai et al. [39] showed that STARD3 contributes to proliferation in Michigan Cancer Foundation-7 (MCF7) cells and adhesion in Monroe Dunaway Anderson- Metastatic Breast-231 (MDA-MB-231) cells. This study pointed to a function of STARD3 in cell matrix adhesion and suggested a regulatory role for STARD3 on focal adhesion kinase (FAK; [39]). These cell models are, however, representative of luminal and basal breast subtypes and therefore, whether these functions are relevant to the HER2 subtype remains to be addressed. In HER2-amplified cells, Sahlberg et al. [44] did a systematic study of HER2 co-amplified genes by LOF and found that several genes of the amplicon, including STARD3 and GRB7, decreased cell viability. They also found that silencing of both STARD3 and HER2 had an additive effect on decreased cell viability and increased apoptosis [44]. These authors evoke the concept of "oncogene addiction" to explain this phenotype. This concept characterizes the dependency of some cancers on one or a few genes for the maintenance of the malignant phenotype [45].

In conclusion, in cancer STARD3 is linked at the genetic and functional levels with HER2. HER2-amplified cancer cells are dependent on co-amplified genes including STARD3 for their growth and survival. In contrast, in other cancer cells STARD3 is dispensable for growth. Actually, STARD3-deficient mice are viable and do not show any growth defects indicating that STARD3 is not essential for cell survival under normal conditions [46]. Paradoxically, STARD3-forced expression is toxic in cancer cells which are not HER2-amplified. Moreover, acute STARD3 expression in the liver of mice induced damage and apoptosis [47]. Cancer is a multistage process, and cancer cells (particularly HER2-amplified cancer cells) have unstable genetic material and they must continuously adapt to maintain their

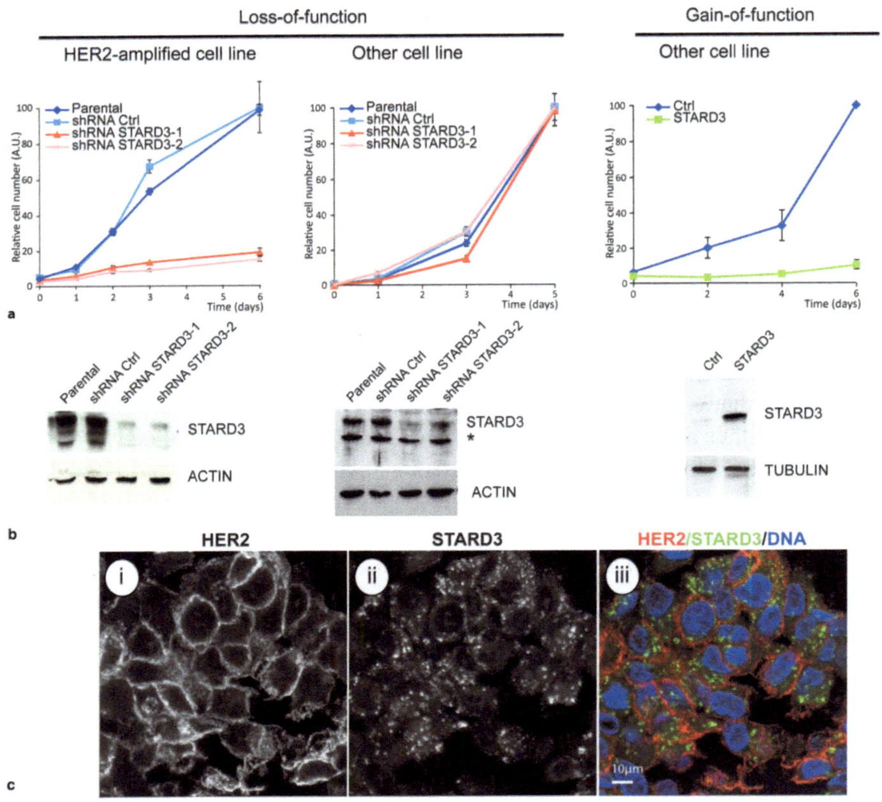

Fig. 6.4 HER2-amplified breast cancer cells are addicted to STARD3. **a** Growth *curves* illustrating the paradoxical role of STARD3 on cell proliferation. Cell growth was analyzed using the 3-(4,5-dimethylthiazol-2-yl)-2,5-diphenyltetrazolium bromide (MTT) assay. *Left*: growth *curves* for the breast cancer HER2-amplified cell line SK-BR-3 after STARD3 silencing. Compared to the parental and control shRNA *cell lines*, clones silenced for STARD3 using two distinct silencing constructs have an impaired growth. *Middle*: cell growth *curve* for HeLa cells after STARD3 silencing. The same experiment was repeated in a HER2-non amplified *cell line*. In HeLa cells, (*cervical carcinoma cells*), silencing of STARD3 using the same shRNA strategy does not compromise cell growth. *Right*: cell growth *curve* after STARD3 overexpression in HeLa cells. HeLa cells, expressing high levels of STARD3 have an impaired growth compared to control (*empty vector*) cells. **b** STARD3 expression in the *cell lines* analyzed in **a**. Protein extracts (~ 10 µg) were analyzed by immunoblotting using anti-STARD3-, -Actin-, and –Tubulin-specific antibodies. In the middle panel a long exposure time was necessary to detect endogenous STARD3 expression in HeLa cells, the *asterisk* represents a non-specific band. **c** HER2 and STARD3 do not co-localize in an invasive ductal carcinoma of the breast. Double immunofluorescence in the same tissue section shows: (i) HER2-plasma membrane staining; (ii) STARD3, cytoplasmic granular staining; (iii) merge image of HER2 (*red*), STARD3 (*green*) and nucleus (*blue*); note the absence of *yellow* staining revealing the lack of co-localization. *HER2* epidermal growth factor receptor 2, *shRNA* short hairpin ribonucleic acid

integrity, survive, and proliferate. The "oncogene addiction" concept suggests that signal transduction and gene expression are regulated in very different manners in cancer cells compared to normal cells, making the former more dependent on the activity of specific genes [45]. We propose that HER2-amplified breast cancers have evolved in a way that they are dependent on STARD3. To date, the molecular details of this dependency are unclear, but STARD3 likely represents an additional therapeutic target in HER2-amplified cancers.

II. STARD3, a Carotenoid-Binding Protein

An unexpected function for STARD3 was revealed in silkworms (Bombyx mori). Silkworm larvae build cocoons which have a yellow color resulting from the presence of carotenoids [48]. Interestingly, animals cannot synthetize carotenoids *de novo* which means that silk carotenoids have a dietary origin, namely mulberry leaves. Therefore, carotenoids have to be absorbed by the gut, transported through the hemolymph to finally be delivered in the silk gland of the larvae where they will give the silk its color. This implies that there are transport mechanisms allowing carotenoids to be conveyed across epithelial cells from the gut and the silk gland to reach their final destination. Between the gut and the silk gland, carotenoids are transported within the hemolymph by lipoprotein particles named lipophorins. The color of carotenoids allowed for the purification of a carotenoid-binding protein, termed CBP, from larval silk glands that appeared yellow under non-denaturing conditions [49]. The main carotenoid bound to CBP was lutein, with β-carotene and α-carotene being less frequently found. The molecular characterization of CBP cDNA showed that it belongs to the START protein family and is closely related to STARD1 and STARD3. However, unlike STARD1 and STARD3 which possess a mitochondria- and an endosome-targeting signal, respectively, CBP is a START-only protein. Intriguingly, the gene encoding CBP uses alternative promoter usage and splicing to produce an isoform closely resembling STARD3; indeed, this CBP isoform termed (BmStart1) possesses a MENTAL domain preceding the START domain [50]. Several silkworm mutants producing white cocoons have been collected over the years of sericulture. For instance, mutant silkworms for the *Y* (*Yellow blood*) gene display colorless hemolymph and silk resulting from an inefficient carotenoid absorption in the gut. Interestingly, in the *Y*-recessive strain, the locus encoding BmStart1/CBP is altered due to a retrotransposon insertion in a CBP-specific exon [51]. As a consequence, while BmStart1 protein is normally present in this strain, CBP is absent, thus suggesting that it is the CBP isoform that is responsible for the silk color. Accordingly, transgenic re-expression of CBP in the *Y* mutant strain restores a yellow color to the silk.

To conclude, the STARD3 gene has evolved in silkworms to produce two different proteins, CBP which is involved in carotenoid transport and consequently silk color and BmStart1, which is similar to STARD3 found in other animals and is likely involved in cholesterol transport. In humans, lutein and zeaxanthin are two carotenoids that accumulate specifically in the central region of the retina named the macula lutea, the part of the retina that is responsible for central vision and is named for its yellow color resulting from the presence of carotenoids. The specific accumulation of these lipophilic molecules in the retina is most probably induced by the action of a molecular transporter. Accordingly, a carotenoid-binding protein was purified from human retina [52]. This protein of ~ 50 kDa binds lutein with a Kd of 0.45 μM and is detected with antibodies raised against the silkworm CBP. Considering the molecular weight of this transporter and its recognition by the anti-CBP antibodies [52], these data suggest that the protein involved in lutein accumulation in human retina is STARD3. In accordance with this idea, the recombinant START domain of STARD3 binds lutein with a Kd of 0.45 μM [53]. These data raise an intriguing question: Is STARD3 responsible for lutein accumulation in the human retina in a manner similar to the silk-worm ortholog allowing lutein deposition in silk? This would imply that the START domain of STARD3 can bind different ligands; it may have a specialized function in lutein handling in some organs and a more general function in cholesterol transport in most cells.

III. STARD3 and Cholesterol Accumulation Disorders

STARD3 as a Surrogate for STARD1-Independent Steroidogenesis STARD1 plays a central role in *de novo* steroidogenesis, however, STARD1 is not expressed in some steroidogenic organs such as the placenta and was long thought to be absent in the brain, suggesting that an alternative cholesterol transport pathway into the mitochondria might exist. Early on, STARD3 was proposed to act as a STARD1-like protein in the placenta and the brain and many lines of evidence supported this hypothesis. STARD3 is expressed in the placenta and the brain [2, 6, 54]. In isolation, the START domain of STARD3 has a StAR-like activity in in vitro steroidogenic assays [6, 16]. Consistent with its tight attachment to the LEs membrane, the full-length protein is, however, poorly active in steroidogenic assays [6, 15, 17]. Of interest, in the placenta, a truncated STARD3 protein leading to the release of an active START domain of about 30 kDa has been described by several groups [6, 17, 55]. The presence of a cleaved form corresponding to the isolated STARD3-START domain supports the notion that STARD3 might act in placental steroidogenesis. However, the generation of mouse models deficient for STARD3 did not substantiate its role in steroidogenesis, at least in the mouse. Indeed, mice lacking STARD3 appear normal and show no reproductive defects [46]. However, the placenta can make estrogens from an alternative pathway which is not initiated by cholesterol conversion into pregnenolone in the mitochondria. The placenta can use steroids produced by the fetal adrenal and convert them to estrogens [56]. Whether this pathway is favored in STARD3-deficient mice remains to be addressed. Alternatively, other START domain containing proteins like the widely expressed STARD4 and STARD5 proteins might compensate for the lack of STARD3. Besides the placenta, the gonads, and the adrenal glands, steroidogenesis occurs in the brain. While STARD3 is well expressed in the brain, both in neurons and in glial cells, it does not co-localize with P450scc which makes it unlikely to be involved in neurosteroid synthesis [54, 57].

STARD3 in Niemann-Pick Type C-Deficient Cells The position of STARD3 at the surface of LEs suggests that it contributes to the distribution of cholesterol from this organelle. Besides *de novo* synthesis in the ER, cells can obtain cholesterol from circulating LDL. The ApoB protein component of LDL binds to LDL receptor family members, the ligand-receptor complex is taken up by clathrin-mediated endocytosis and is dissociated in endosomes. While the receptor is recycled back to the cell surface, the lipid component of LDL is targeted to LEs/Lysosomes (LE/Ly) where cholesterol becomes redistributed inside the cell [58]. How exactly cholesterol is trafficked from endosomes to distinct intracellular regions remains unclear. Studies of Niemann-Pick type C disease, a lipid storage disorder which leads to neurological disorders and hepatosplenomegaly (reviewed in [59–61]), revealed the instrumental role of two proteins, Niemann-Pick C1 (NPC1) and Niemann-Pick C2 (NPC2), on cholesterol egress from LE/Ly. NPC proteins are both LE resident proteins acting in concert; NPC2, a soluble LE/Ly luminal protein binds cholesterol in the lumen, and exchanges cholesterol with the N-terminal cholesterol-binding domain of NPC1 [62]. To address the potential role of STARD3 in cholesterol clearing from LE, STARD3 was overexpressed in NPC-deficient fetal fibroblasts but no correction of the cholesterol accumulation phenotype was observed [12, 63]. Mechanistic insights explaining the lack of connection between the NPC proteins and STARD3 were proposed recently [64]. According to this study, endosomal cholesterol would be trafficked in a sequential manner, first from LE to the plasma membrane through endosomes containing STARD3 and the cholesterol transporter ABCA3 and second from LE to the ER through a distinct set of endosomes positive for the oxysterol-binding protein-related protein 1L (ORP1L) and NPC1 [64]. This model is notably based on the finding that STARD3 and NPC1 mark two distinct endosome pools and is consistent with the fact that overexpression of STARD3 does not significantly increase ACAT-mediated cholesterol esterification in the ER [23]. Further, STARD3 overexpression mimics some of the features of NPC-deficient cells. In all cell types tested STARD3 overexpression was associated with sterol deposition in LE [12, 23, 63]. Moreover, acute overexpression of STARD3 in mouse liver was associated with increased cholesterol content, biliary bile acid concentration, and impaired bile flow and caused severe liver damage and apoptosis[47]. All these studies argue against a role of STARD3 in cholesterol egress from LE/Ly. Other studies have, however, provided evidence that STARD3 promotes cholesterol transfer presumably from LDL to the mitochondria. Charman et al. [65] showed an increase transport of cholesterol to mitochondria in NPC1-deficient Chinese Hamster ovary (CHO) cells that is mediated by STARD3. Consistent with this, Zhang et al. [66] proposed the involvement of STARD3 in the transfer of cholesterol to the mitochondria. Indeed, using time lapse imaging techniques, these authors found that STARD3 LE tubules aligned parallel to mitochondria and reported the presence of transient contact between these two organelles. The fact that the efforts made to clarify the function of STARD3 in cholesterol trafficking have used many cancer cells models may explain some of the discrepancies found in the literature.

While the role of STARD3 in cholesterol transfer from LE is still unclear, our new finding that STARD3-positive endosomes are tethered with the ER, argues for its participation in cholesterol egress from LE to the ER [26]. However, experimental data do not support this model. The cholesterol transfer function of STARD3 appears to be different from this tethering activity. The trafficking of endosomal cholesterol is probably complex and sequential, it may be regulated by redundant mechanisms, is probably cell-type specific and may be altered differently in cancer cells. In the future, mechanistic analysis of the function of STARD3 in cholesterol trafficking should use both in vitro and in vivo models that are physiological and highly relevant to human disease.

Summary

The cloning of STARD3 was fairly rapid and easy; the functional characterization of this protein has been difficult and is far from being achieved. To date, an understanding of how STARD3 handles cholesterol has been elusive. From its homology with STARD1, it was proposed that STARD3 functions to transport cholesterol from endosomes to the ER. Many studies including those using NPC-deficient cells and the generation of STARD3-deficient mice did not support this hypothesis and it is clear that additional studies are necessary to uncover the role of STARD3 in cholesterol trafficking. STARD3 is overexpressed in the HER2-subtype of breast cancer, it belongs to the smallest region of amplification involving HER2 and its presence in breast cancer is associated with a poor prognosis. Consistent with this, STARD3-silencing results in cell growth arrest and apoptosis in HER2-positive cell lines. How STARD3 contributes to the phenotype of HER2 cancers is still unclear. Unlike STARD1, STARD3 is not essential for steroidogenesis; however, STARD3 is a lipid-binding protein involved in cholesterol trafficking at the level of LEs. Remarkably, STARD3 modifies the intracellular distribution and morphology of LEs and moreover STARD3 forms membrane contact sites between LEs and the endoplasmic reticulum. These organelles are central for cholesterol homeostasis. The ER is the site of cholesterol biosynthesis while LEs ensure LDL-derived cholesterol distribution. Thus, the physical location of STARD3 between these two organelles places the protein at a unique position at the interface of two major cholesterol pathways. Moreover, extended ER-endosome contacts regulate the dynamics of the endocytic compartment probably by reducing the maturation of LEs to lysosomes. As the ER and endosomes are the sites of synthesis and degradation of membrane-receptors, respectively, we can speculate that STARD3 acts on cancer cells by modifying the lipid composition of cellular membranes. Indeed, membranes are not only barriers, they also function as interfaces at which numerous cellular processes, including signaling and cell death mechanisms, are concentrated and regulated. Despite all these uncertainties, we have made considerable progress in clarifying the function of STARD3. It is without doubt that this fascinating protein will keep us busy for the coming years.

References

1. Tomasetto C, Regnier C, Moog-Lutz C, Mattei MG, Chenard MP, Lidereau R, et al. Identification of four novel human genes amplified and overexpressed in breast carcinoma and localized to the q11-q21.3 region of chromosome 17. Genomics. 1995 Aug 10;28(3):367–76.
2. Moog-Lutz C, Tomasetto C, Regnier CH, Wendling C, Lutz Y, Muller D, et al. MLN64 exhibits homology with the steroidogenic acute regulatory protein (STAR) and is over-expressed in human breast carcinomas. Int J Cancer. 1997 April 10;71(2):183–91.
3. Alpy F, Tomasetto C. MLN64 and MENTHO, two mediators of endosomal cholesterol transport. Biochem Soc Trans. 2006 June;34(Pt 3):343–5.
4. Bieche I, Tomasetto C, Regnier CH, Moog-Lutz C, Rio MC, Lidereau R. Two distinct amplified regions at 17q11-q21 involved in human primary breast cancer. Cancer Res. 1996 Sept 1;56(17):3886–90.
5. Akiyama N, Sasaki H, Ishizuka T, Kishi T, Sakamoto H, Onda M, et al. Isolation of a candidate gene, CAB1, for cholesterol transport to mitochondria from the c-ERBB-2 amplicon by a modified cDNA selection method. Cancer Res. 1997 Aug 15;57(16):3548–53.
6. Watari H, Arakane F, Moog-Lutz C, Kallen CB, Tomasetto C, Gerton GL, et al. MLN64 contains a domain with homology to the steroidogenic acute regulatory protein (StAR) that stimulates steroidogenesis. Proc Natl Acad Sci U S A. 1997 Aug 5;94(16):8462–7.
7. Ponting CP, Aravind L. START: a lipid-binding domain in StAR, HD-ZIP and signalling proteins. Trends Biochem Sci. 1999 April;24(4):130–2.
8. Soccio RE, Breslow JL. StAR-related lipid transfer (START) proteins: mediators of intracellular lipid metabolism. J Biol Chem. 2003 June 20;278(25):22183–6.
9. Alpy F, Tomasetto C. Give lipids a START: the StAR-related lipid transfer (START) domain in mammals. J Cell Sci. 2005 July 1;118(Pt 13):2791–801.
10. Clark BJ. The mammalian START domain protein family in lipid transport in health and disease. J Endocrinol. 2012 March;212(3):257–75.
11. Alpy F, Tomasetto C. START ships lipids across interorganelle space. Biochimie. 2013 Sept 25;96:85–95.
12. Alpy F, Stoeckel ME, Dierich A, Escola JM, Wendling C, Chenard MP, et al. The steroidogenic acute regulatory protein homolog MLN64, a late endosomal cholesterol-binding protein. J Biol Chem. 2001 Feb 9;276(6):4261–9.
13. Alpy F, Wendling C, Rio MC, Tomasetto C. MENTHO, a MLN64 homologue devoid of the START domain. J Biol Chem. 2002 Dec 27;277(52):50780–7.
14. Alpy F, Latchumanan VK, Kedinger V, Janoshazi A, Thiele C, Wendling C, et al. Functional characterization of the MENTAL domain. J Biol Chem. 2005 May 6;280(18):17945–52.
15. Tuckey RC, Bose HS, Czerwionka I, Miller WL. Molten globule structure and steroidogenic activity of N-218 MLN64 in human placental mitochondria. Endocrinology. 2004 April;145(4):1700–7.
16. Bose HS, Baldwin MA, Miller WL. Evidence that StAR and MLN64 act on the outer mitochondrial membrane as molten globules. Endocr Res. 2000 Nov;26(4):629–37.
17. Bose HS, Whittal RM, Huang MC, Baldwin MA, Miller WL. N-218 MLN64, a protein with StAR-like steroidogenic activity, is folded and cleaved similarly to StAR. Biochemistry. 2000 Sept 26;39(38):11722–31.
18. Bose HS, Whittal RM, Ran Y, Bose M, Baker BY, Miller WL. StAR-like activity and molten globule behavior of StARD6, a male germ-line protein. Biochemistry. 2008 Feb 26;47(8):2277–88.
19. Tsujishita Y, Hurley JH. Structure and lipid transport mechanism of a StAR-related domain. Nat Struct Biol. 2000 May;7(5):408–14.
20. Shalchian-Tabrizi K, Minge MA, Espelund M, Orr R, Ruden T, Jakobsen KS, et al. Multigene phylogeny of choanozoa and the origin of animals. PLoS One. 2008;3(5):e2098.
21. King N, Westbrook MJ, Young SL, Kuo A, Abedin M, Chapman J, et al. The genome of the choanoflagellate Monosiga brevicollis and the origin of metazoans. Nature. 2008 Feb 14;451(7180):783–8.

22. Alpy F, Boulay A, Moog-Lutz C, Andarawewa KL, Degot S, Stoll I, et al. Metastatic lymph node 64 (MLN64), a gene overexpressed in breast cancers, is regulated by Sp/KLF transcription factors. Oncogene. 2003 June 12;22(24):3770–80.
23. Liapis A, Chen FW, Davies JP, Wang R, Ioannou YA. MLN64 transport to the late endosome is regulated by binding to 14-3-3 via a non-canonical binding site. PLoS One. 2012;7(4):e34424.
24. Le Naour F, Andre M, Boucheix C, Rubinstein E. Membrane microdomains and proteomics: lessons from tetraspanin microdomains and comparison with lipid rafts. Proteomics. 2006 Dec;6(24):6447–54.
25. Hulce JJ, Cognetta AB, Niphakis MJ, Tully SE, Cravatt BF. Proteome-wide mapping of cholesterol-interacting proteins in mammalian cells. Nat Methods. 2013 March;10(3):259–64.
26. Alpy F, Rousseau A, Schwab Y, Legueux F, Stoll I, Wendling C, et al. STARD3/STARD3NL and VAP make a novel molecular tether between late endosomes and the ER. J Cell Sci. 2013 Oct 8;126(Pt 23):5500–12.
27. Loewen CJ, Roy A, Levine TP. A conserved ER targeting motif in three families of lipid binding proteins and in Opi1p binds VAP. EMBO J. 2003 May 1;22(9):2025–35.
28. Polyak K, Metzger Filho O. SnapShot: breast cancer. Cancer Cell. 2012 Oct 16;22(4):562–e1.
29. Higgins MJ, Baselga J. Targeted therapies for breast cancer. J Clin Invest. 2011 Oct;121(10):3797–803.
30. Ellsworth RE, Ellsworth DL, Patney HL, Deyarmin B, Love B, Hooke JA, et al. Amplification of HER2 is a marker for global genomic instability. BMC Cancer. 2008;8:297.
31. Jacot W, Fiche M, Zaman K, Wolfer A, Lamy PJ. The HER2 amplicon in breast cancer: Topoisomerase IIA and beyond. Biochim Biophys Acta. 2013 Aug;1836(1):146–57.
32. Perou CM, Sorlie T, Eisen MB, van de Rijn M, Jeffrey SS, Rees CA, et al. Molecular portraits of human breast tumours. Nature. 2000 Aug 17;406(6797):747–52.
33. Pollack JR, Perou CM, Alizadeh AA, Eisen MB, Pergamenschikov A, Williams CF, et al. Genome-wide analysis of DNA copy-number changes using cDNA microarrays. Nat Genet. 1999 Sept;23(1):41–6.
34. Sorlie T, Perou CM, Tibshirani R, Aas T, Geisler S, Johnsen H, et al. Gene expression patterns of breast carcinomas distinguish tumor subclasses with clinical implications. Proc Natl Acad Sci U S A. 2001 Sept 11;98(19):10869–74.
35. Kauraniemi P, Barlund M, Monni O, Kallioniemi A. New amplified and highly expressed genes discovered in the ERBB2 amplicon in breast cancer by cDNA microarrays. Cancer Res. 2001 Nov 15;61(22):8235–40.
36. Orsetti B, Nugoli M, Cervera N, Lasorsa L, Chuchana P, Ursule L, et al. Genomic and expression profiling of chromosome 17 in breast cancer reveals complex patterns of alterations and novel candidate genes. Cancer Res. 2004 Sept 15;64(18):6453–60.
37. Vincent-Salomon A, Lucchesi C, Gruel N, Raynal V, Pierron G, Goudefroye R, et al. Integrated genomic and transcriptomic analysis of ductal carcinoma in situ of the breast. Clin Cancer Res. 2008 April 1;14(7):1956–65.
38. Vinatzer U, Dampier B, Streubel B, Pacher M, Seewald MJ, Stratowa C, et al. Expression of HER2 and the coamplified genes GRB7 and MLN64 in human breast cancer: quantitative real-time reverse transcription-PCR as a diagnostic alternative to immunohistochemistry and fluorescence in situ hybridization. Clin Cancer Res. 2005 Dec 1;11(23):8348–57.
39. Cai W, Ye L, Sun J, Mansel RE, Jiang WG. Expression of MLN64 influences cellular matrix adhesion of breast cancer cells, the role for focal adhesion kinase. Int J Mol Med. 2010 April;25(4):573–80.
40. Lamy PJ, Fina F, Bascoul-Mollevi C, Laberenne AC, Martin PM, Ouafik L, et al. Quantification and clinical relevance of gene amplification at chromosome 17q12-q21 in human epidermal growth factor receptor 2-amplified breast cancers. Breast Cancer Res. 2011;13(1):R15.
41. Kao J, Pollack JR. RNA interference-based functional dissection of the 17q12 amplicon in breast cancer reveals contribution of coamplified genes. Genes Chromosomes Cancer. 2006 Aug;45(8):761–9.
42. Hynes NE, MacDonald G. ErbB receptors and signaling pathways in cancer. Curr Opin Cell Biol. 2009 April;21(2):177–84.

43. Baulida J, Kraus MH, Alimandi M, Di Fiore PP, Carpenter G. All ErbB receptors other than the epidermal growth factor receptor are endocytosis impaired. J Biol Chem. 1996 March 1;271(9):5251–7.
44. Sahlberg KK, Hongisto V, Edgren H, Makela R, Hellstrom K, Due EU, et al. The HER2 amplicon includes several genes required for the growth and survival of HER2 positive breast cancer cells. Mol Oncol. 2013 June;7(3):392–401.
45. Weinstein IB. Cancer. Addiction to oncogenes-the Achilles heal of cancer. Science. 2002 July 5;297(5578):63–4.
46. Kishida T, Kostetskii I, Zhang Z, Martinez F, Liu P, Walkley SU, et al. Targeted mutation of the MLN64 START domain causes only modest alterations in cellular sterol metabolism. J Biol Chem. 2004 April 30;279(18):19276–85.
47. Tichauer JE, Morales MG, Amigo L, Galdames L, Klein A, Quinones V, et al. Overexpression of the cholesterol-binding protein MLN64 induces liver damage in the mouse. World J Gastroenterol. 2007 June 14;13(22):3071–9.
48. Sakudoh T, Tsuchida K. Chapter 24: transport of carotenoids by a carotenoid-binding protein in the silkworm. In: Landrum JT, editor. Carotenoids: physical, chemical, and biological functions and properties. Boca Raton: CRC Press; 2009.
49. Tabunoki H, Sugiyama H, Tanaka Y, Fujii H, Banno Y, Jouni ZE, et al. Isolation, characterization, and cDNA sequence of a carotenoid binding protein from the silk gland of Bombyx mori larvae. J Biol Chem. 2002 Aug 30;277(35):32133–40.
50. Sakudoh T, Tsuchida K, Kataoka H. BmStart1, a novel carotenoid-binding protein isoform from Bombyx mori, is orthologous to MLN64, a mammalian cholesterol transporter. Biochem Bioph Res Co. 2005 Nov 4;336(4):1125–35.
51. Sakudoh T, Sezutsu H, Nakashima T, Kobayashi I, Fujimoto H, Uchino K, et al. Carotenoid silk coloration is controlled by a carotenoid-binding protein, a product of the Yellow blood gene. Proc Natl Acad Sci U S A. 2007 May 22;104(21):8941–6.
52. Bhosale P, Li B, Sharifzadeh M, Gellermann W, Frederick JM, Tsuchida K, et al. Purification and partial characterization of a lutein-binding protein from human retina. Biochemistry. 2009 June 9;48(22):4798–807.
53. Li B, Vachali P, Frederick JM, Bernstein PS. Identification of StARD3 as a lutein-binding protein in the macula of the primate retina. Biochemistry. 2011 April 5;50(13):2541–9.
54. King SR, Smith AG, Alpy F, Tomasetto C, Ginsberg SD, Lamb DJ. Characterization of the putative cholesterol transport protein metastatic lymph node 64 in the brain. Neuroscience. 2006;139(3):1031–8.
55. Olvera-Sanchez S, Espinosa-Garcia MT, Monreal J, Flores-Herrera O, Martinez F. Mitochondrial heat shock protein participates in placental steroidogenesis. Placenta. 2011 March;32(3):222–9.
56. Miller WL. Steroid hormone synthesis in mitochondria. Mol Cell Endocrinol. 2013 Oct 15;379(1-2):62–73.
57. King SR, Ginsberg SD, Ishii T, Smith RG, Parker KL, Lamb DJ. The steroidogenic acute regulatory protein is expressed in steroidogenic cells of the day-old brain. Endocrinology. 2004 Oct;145(10):4775–80.
58. Goldstein JL, Brown MS. The LDL receptor. Arterioscler Thromb Vasc Biol. 2009 April;29(4):431–8.
59. Vanier MT, Millat G. Niemann-Pick disease type C. Clin Genet. 2003 Oct;64(4):269–81.
60. Liscum L, Sturley SL. Intracellular trafficking of Niemann-Pick C proteins 1 and 2: obligate components of subcellular lipid transport. Biochim Biophys Acta. 2004 Oct 11;1685(1–3):22–7.
61. Sturley SL, Patterson MC, Balch W, Liscum L. The pathophysiology and mechanisms of NP-C disease. Biochim Biophys Acta. 2004 Oct 11;1685(1–3):83–7.
62. Deffieu MS, Pfeffer SR. Niemann-Pick type C 1 function requires lumenal domain residues that mediate cholesterol-dependent NPC2 binding. Proc Natl Acad Sci U S A. 2011 Nov 22;108(47):18932–6.

63. Holtta-Vuori M, Alpy F, Tanhuanpaa K, Jokitalo E, Mutka AL, Ikonen E. MLN64 is involved in actin-mediated dynamics of late endocytic organelles. Mol Biol Cell. 2005 Aug;16(8):3873–86.
64. van der Kant R, Zondervan I, Janssen L, Neefjes J. Cholesterol-binding molecules MLN64 and ORP1 Lmark distinct late endosomes with transporters ABCA3 and NPC1. J Lipid Res. 2013 Aug;54(8):2153–65.
65. Charman M, Kennedy BE, Osborne N, Karten B. MLN64 mediates egress of cholesterol from endosomes to mitochondria in the absence of functional Niemann-Pick Type C1 protein. J Lipid Res. 2010 May;51(5):1023–34.
66. Zhang M, Liu P, Dwyer NK, Christenson LK, Fujimoto T, Martinez F, et al. MLN64 mediates mobilization of lysosomal cholesterol to steroidogenic mitochondria. J Biol Chem. 2002 Sept 6;277(36):33300–10.
67. Letunic I, Bork P. Interactive Tree of Life v2: online annotation and display of phylogenetic trees made easy. Nucleic Acids Res. 2011 July;39(Web Server issue):W475–8.

Chapter 7
The STARD4 Subfamily: STARD4 and STARD5 in Cholesterol Metabolism

Raymond E. Soccio

Abstract The STARD4 subfamily of steroidogenic acute regulatory protein (StAR)-related lipid transfer (START) proteins consists of STARD4, STARD5, and STARD6, which share ~30 % amino acid identity with each other, ~20 % with the StAR/STARD1 and STARD3/metastatic lymph node clone 64 (MLN64) START domains, and ~15 % with other START domains. All three have no other protein domains besides START. Since their initial discovery, they were proposed to serve as intracellular sterol transporters. A subsequent decade of research has led to seven high-confidence conclusions: (1) STARD4 expression is regulated by cellular sterol levels via the sterol regulatory element binding protein (SREBP)-2 transcription factor. (2) Endoplasmic reticulum (ER) stress-inducing agents increase STARD5 levels, though the mechanism may be posttranscriptional. (3) STARD4 and STARD5 are cytosolic proteins that may associate loosely with specific subcellular membranes. (4) STARD4 can bind cholesterol and efficiently transfer it between membranes. (5) STARD4 can mediate transfer of cholesterol to the ER resident enzyme acyl-coenzyme A cholesterol acyltransferase (ACAT) for esterification. (6) STARD5 does not have these ACAT effects but has some distinct activities from STARD4. (7) STARD4 null mice lack any clear phenotype. Other conclusions are less certain due to conflicting data but merit further study: ER stress and steroidogenic regulation of STARD4, selective expression of STARD5 in immune and reticuloendothelial cells, nuclear localization of STARD4 and STARD5, binding of bile acids by STARD5, and binding of other sterols besides cholesterol by STARD4 (such as 7α-hydroxycholesterol) and STARD5 (such as 25-hydroxycholesterol). Selective expression of STARD6 in male germ cells strongly suggests a role in fertility, but the functions of STARD4 and STARD5 in normal physiology and disease remain elusive.

List of Abbreviations

ACAT	Acyl-coenzyme A Cholesterol Acyltransferase
ATF6	Activating Transcription Factor 6
CD	Circular Dichroism

R. E. Soccio, M.D., Ph.D. (✉)
Division of Endocrinology, Diabetes, and Metabolism, Smilow Center for Translational Research, Perelman School of Medicine, University of Pennsylvania,
12th floor, 3400 Civic Center Blvd, Philadelphia, PA 19104, USA
e-mail: soccio@mail.med.upenn.edu

B. J. Clark, D. M. Stocco (eds.), *Cholesterol Transporters of the START Domain Protein Family in Health and Disease,* DOI 10.1007/978-1-4939-1112-7_7,

ChIP-seq	Chromatin Immunoprecipitation Followed by Deep Sequencing
CHOP	CCAAT/Enhancer-binding Protein Homologous Protein
DHE	Dihydroergosterol
ER	Endoplasmic Reticulum
ERC	Endocytic Recycling Compartment
EST	Expressed Sequence Tag
FRAP	Fluorescence Recovery after Photobleaching
GRP94	Glucose-regulated Protein 94
Herp	Homocysteine-induced Endoplasmic Reticulum Protein
HMGCR	HMG CoA (3-hydroxy-3-methylglutaryl coenzyme A) Reductase
INSIG	Insulin-induced Gene
LDL	Low-density Lipoprotein
MCD	Metylcyclodextrin
NMR	Nuclear Magnetic Resonance
SCAP	SREBP Cleavage-activating Protein
SREBP	Sterol Regulatory Element Binding Protein
StAR	Steroidogenic Acute Regulatory Protein
STARD4	START Domain Containing 4
STARD5	START Domain Containing 5
STARD6	START Domain Containing 6
START	StAR-related Lipid Transfer
UTR	Untranslated Region
Xbp1	X-box Binding Protein 1

Introduction: A Personal Account of Discovery

I was an MD-PhD graduate student at Rockefeller University whose primary project was floundering, so my secondary project was a fishing expedition. Our overarching goal was to address the question of how the body responds to dietary cholesterol, such that some individuals are sensitive and increase plasma cholesterol while others are insensitive. Genes in the liver that change with dietary cholesterol seemed like a good place to start, so we fed mice a standard diet (0.02% cholesterol) versus a high cholesterol diet (0.5%) for three weeks. At the time around the year 2000, microarrays to measure messenger ribonucleic acid (mRNA) gene expression were the latest in technology. The wide availability of commercial arrays was a year or so away, so we did not have access to the familiar oligonucleotide probe-based systems like Affymetrix GeneChips, which are now being replaced by RNA-sequencing. Instead, we used complementary deoxyribonucleic acid (cDNA) arrays in which entire cDNA clones (each hundreds of base pairs) were spotted onto slides. To obtain and analyze these microarrays, my excellent mentor, Jan Breslow, established a collaboration with Raju Kucherlapati's laboratory, then at Albert Einstein College of Medicine. I remember taking the subway from Manhattan to the Bronx to see the wondrous custom-built robotic system that spotted approximately 9,000 cDNA clones from 96-well plates onto glass slides. About half of these cDNAs were annotated with gene names, while the rest were unidentified expressed sequence tags (ESTs).

In our initial microarray hybridizations, we noted marked variation across experiments, as the statistical tools to analyze microarray data were also in their infancy. After multiple biological and technical replicates and stringent cutoffs, only six transcripts were consistently downregulated more than twofold by dietary cholesterol, and disappointingly none were convincingly upregulated. Identifying the genes corresponding to six regulated transcripts was not trivial given that the annotation of the mouse genome was far from complete. Five of the six genes turned out to encode enzymes involved in cholesterol or fatty acid synthesis. This was an excellent proof of principle for our experimental system, since these pathways were expected to be downregulated by cholesterol, negative feedback mediated by the sterol regulatory element binding protein (SREBP) transcription factors.

The sixth transcript was a mystery. EST AA239481 was cloned from mouse liver, not annotated with any gene name, and the available 460 bases of sequence in the database did not contain any obvious protein-coding region or homology to known genes. We obtained the EST clone and sequenced its entire insert of 1,114 base pairs, but did not find an open reading frame. Since the cDNA was cloned with oligo-dT for the 3′ polyadenylation site, I suspected that this sequence was in the 3′ untranslated region (UTR) of the gene. I was self-taught at using the relatively primitive bioinformatic resources available, and I remember spending a late evening in the lab at my computer staring at this sequence. The breakthrough happened when I performed a basic local alignment search tool (BLAST) search and found overlap with another uncharacterized EST, extending the cDNA sequence by several hundred nucleotides in the presumed 5′ direction but still not revealing the coding sequence. However, this new EST overlapped with another, and then another, so by walking from one EST to the next across about 3 kilobases I finally found protein coding sequence. Ultimately, alignment of EST and genomic sequence revealed an unannotated six exon gene encoding a 224 amino acid protein, with the final exon including more than 4 kilobases of 3′ UTR including the initial EST.

Evening had turned into night and I had completely lost track of time immersed in this DNA sequence analysis, but the biggest surprise came when I searched for known proteins with homology to this novel one. The first hit was something called "StAR (Steroidogenic Acute Regulatory Protein)." I had never heard of it, but I immediately performed literature searches and was thrilled to find an intracellular cholesterol transport protein, essential for delivering cholesterol to mitochondria for steroid hormone synthesis [1]. The StAR-related lipid transfer (START) domain had been described [2] and the crystal structure of the related metastatic lymph node clone 64 (MLN64) START domain was published [3], with the beautiful hydrophobic cavity to bind lipid. My mentor Jan Breslow was instrumental in the molecular cloning of many apolipoprotein genes [4], which are essential for transport of otherwise insoluble cholesterol and other lipids in the blood. Cholesterol of course has the same problem of insolubility within the cell, and here we had found a putative cholesterol transport protein whose expression was regulated by dietary cholesterol! After a sleepless night with excitement, I rushed into Jan's office the next morning to share this result. Jan has a famously calm and even demeanor, but I could see his delight as we allowed ourselves to speculate on the potential functions and implications of a novel sterol transporter.

We decided to name this protein CRSP, pronounced "crisp," for "Cholesterol-Regulated START Protein." With my new experience searching DNA sequence databases, I found and assembled the whole family of 15 mammalian START domain-containing proteins into a phylogenetic tree [5]. Among characterized START proteins, CRSP was most similar to the known cholesterol-binding proteins StAR and MLN64, but there were two other novel START domain proteins even more like CRSP. I creatively called them "CRSP-like1" and "CRSP-like2," and we immediately set out cloning and characterizing these three genes. It quickly became my primary PhD thesis project, displacing studies of the bile acid biosynthetic enzyme cholesterol 7 alpha-hydroxylase (Cyp7A1) [6]. I remember reaching out to other researchers in the START field to obtain reagents and advice, such as Walter Miller, Jerry Strauss, and Doug Stocco. I was amazed how helpful and collaborative they were (in retrospect, with more experience in the competitive world of academic biomedical research, I am only more amazed). In fact, several of them had independently noted the cDNAs for one or more of the three novel genes, but none of them rushed to publish first or compete with us.

When we had assembled enough data for our first publication describing this subfamily of three novel START domain proteins [7], the journal required us to submit to the Human Genome Organisation (HUGO) gene nomenclature committee. My name CRSP was shot down, and instead HUGO decided to rename the whole START domain superfamily with the START domain (Stard/STARD#) nomenclature reflected in the official mouse and human gene symbols today. I objected to these generic names since they fail to describe known physiology, regulation, or function of the proteins, and many of the family members have other protein domains besides a START domain. My objections were to no avail, and CRSP, CRSPL1, and CRSPL2 proteins became STARD4, STARD5, and STARD6, whereas previously named proteins StAR, Phosphatidylcholine transfer protein (PCTP), and MLN64 became STARD1, STARD2, and STARD3, respectively. Please don't blame me for the nomenclature, because I do not like it either!

I finished my PhD work on the STARD4 subfamily, with Fred Maxfield from across the street at Weill Cornell Medical College serving on my thesis committee and advising us on the cell biology of intracellular cholesterol transport. I was never satisfied that we failed to describe a true physiological function for STARD4 and STARD5 in cholesterol metabolism. Two years after my cDNA microarrays, a new and very talented MD-PhD student in the Breslow lab, my good friend Kara Maxwell, performed the same cholesterol feeding experiment yet used Affymetrix microarrays. Of course she found STARD4 again, but she also discovered the proprotein convertase subtilisin/kexin type 9 (Pcsk9) protein [8], which turned out to have a key role in regulating levels of low-density lipoprotein (LDL) "bad" cholesterol and the associated risk of atherosclerosis. Pcsk9 is now an exciting drug target

[9], whereas—despite the diligent efforts of very thorough investigators—STARD4 languishes in relative obscurity. I returned to medical school, then clinical training in internal medicine and endocrinology. This choice of subspecialty was guided by my research interest in lipid metabolism, diabetes, and obesity, but perhaps also a little by the role of STARD1/StAR in steroid hormone synthesis. During my clinical fellowship, my attending physician, Carrie Burns noted a patient with congenital adrenal hyperplasia and unusual biochemical results we interpreted as partial StAR deficiency, which had not been described. Sequencing of his StAR gene indeed revealed compound heterozygosity for two mutations, which we shared with Walter Miller's laboratory for characterization. Along with several other patients, this led to the description of nonclassic/atypical lipoid congenital adrenal hyperplasia due to StAR mutants with partial activity [10]. I had hoped that human disease phenotypes might likewise someday be associated with STARD4 or STARD5, but this has not yet come to pass. My postdoctoral research has taken me away from my beloved START proteins, but I have followed from afar with great interest, and I review the progress here.

Stard4/STARD4 Gene Regulation

As described in detail above, the mouse StarD4 gene was discovered due to its regulation by cholesterol [7]. Cellular cholesterol exerts negative feedback on genes like STARD4 via the SREBP transcription factors, whose mechanism has been elegantly described over the past several decades by the laboratory of Michael Brown and Joseph Goldstein (Fig. 7.1a, reviewed in ref. [11]) Briefly, SREBPs are synthesized as inactive precursors spanning the membrane of the endoplasmic reticulum (ER). When cellular sterols are adequate, SREBP remains in the ER associated with sterol sensing proteins SREBP cleavage-activating protein (SCAP) and Insulin-induced gene (INSIG). Low sterol results in a conformational change in SCAP, releasing INSIG such that SCAP escorts SREBP in vesicles from the ER to the Golgi, where successive proteolytic cleavages of the SREBP transmembrane region by site 1 and site 2 proteases release the soluble N-terminal transcription factor domain. This mature SREBP then translocates to the nucleus and binds to the sterol response elements (SREs) in gene promoters to activate target genes.

Many lines of evidence support the idea that STARD4 is a direct target of SREBP transcription factors, particularly SREBP-2 that predominantly regulates cholesterol metabolism, as opposed to SREBP-1 that predominantly regulates fatty acid metabolism. First, STARD4 is consistently co-regulated by sterols with other classic targets of SREBP-2, such as enzymes in the cholesterol biosynthesis pathway like the rate-limiting enzyme 3-hydroxy-3-methyl-glutaryl (HMG) CoA reductase (HMGCR). In mouse liver, upon 3 weeks of high cholesterol feeding, twofold decreases in STARD4 mRNA are observed on microarrays and validated by quantitative polymerase chain reaction (PCR), coordinately regulated with known SREBP-2 targets [7, 8]. Regulation is even greater in cultured mouse 3T3-L1 cells: sterol

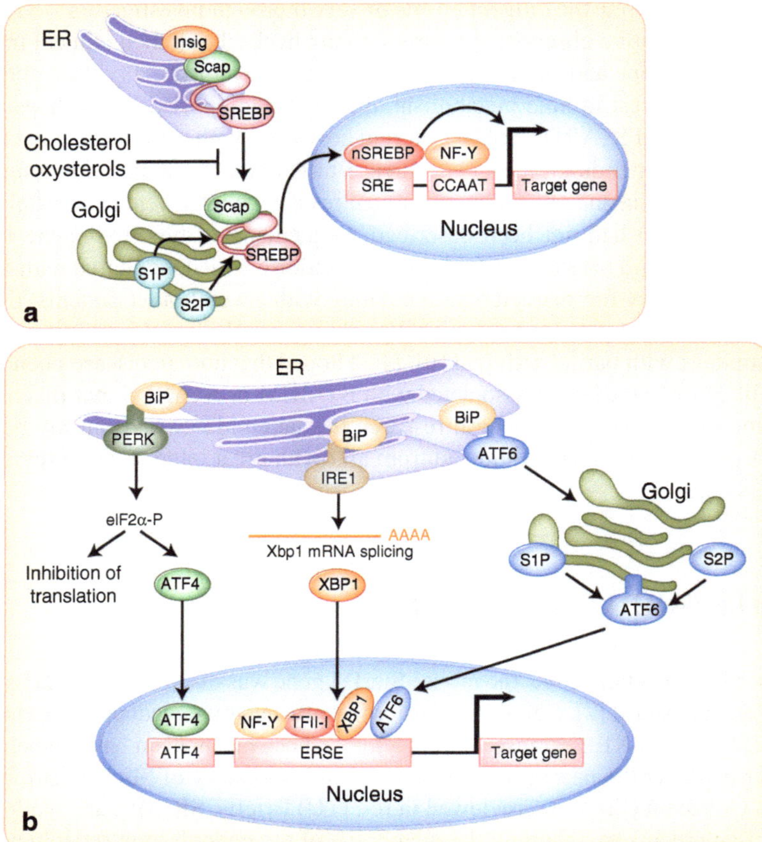

Fig. 7.1 Regulation of SREBP and ER stress transcription factors. **a** SREBPs are activated upon cholesterol depletion. When cellular cholesterol or certain oxysterols are abundant, SREBPs remain in the ER as inactive precursors associated with SCAP and Insig. When sterols are scarce, SCAP releases Insig allowing SREBP-SCAP trafficking to the Golgi apparatus, where the site 1 and site 2 proteases (S1P and S2P) cleave SREBP. This releases the N-terminal transcription factor nSREBP, which translocates to the nucleus and binds promoter SREs. nSREBPs cooperate with other factors like NF-Y to activate transcription of target genes involved in the synthesis and uptake of cholesterol and fatty acids. **b** ER stress signals are transduced to the nucleus via three parallel pathways, activating the ATF4, Xbp1, and ATF6 transcription factors. The RNA-dependent protein kinase-like ER kinase (PERK) pathway transiently inhibits global translation but activates translation of ATF4. IRE1 mediates non-traditional splicing of the Xbp1 mRNA to increase its synthesis, while ATF6 is activated by proteolysis in the Golgi similar to SREBP processing. The ER chaperone BiP negatively regulates all three pathways, which may converge in the nucleus with complex interactions on target gene promoters with ER stress response elements (ERSEs). *SREBP* sterol regulatory element binding protein, *ER* endoplasmic reticulum, *SCAP* SREBP cleavage-activating protein, *ATF* activating transcription factor

depletion with lovastatin to activate SREBP increased STARD4 expression, while addition of 25-hydroxycholesterol to repress SREBP decreased STARD4 expression, with a difference of 14-fold between the conditions [7]. The same sterol regu-

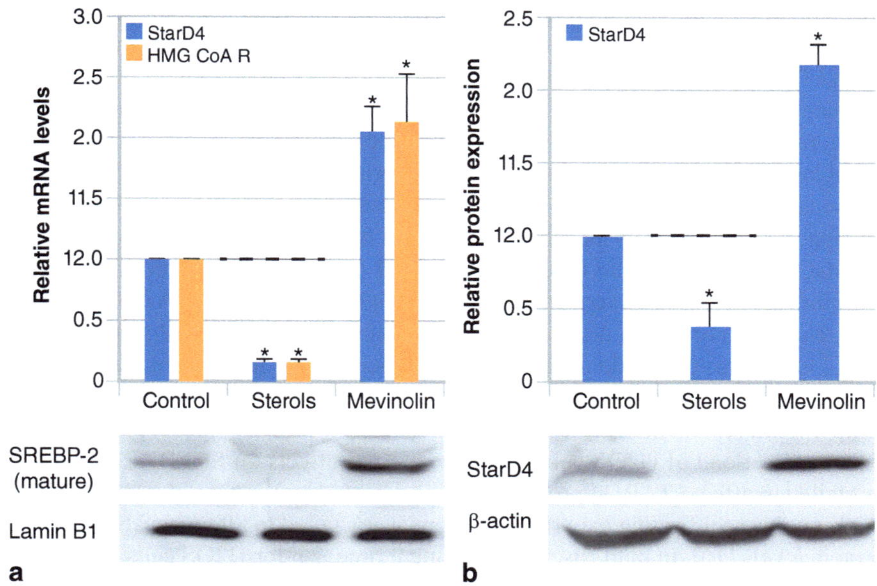

Fig. 7.2 STARD4 expression is regulated by intracellular sterol levels via SREBP-2. **a** In 3T3-L1 fibroblasts, treatment with sterols represses SREBP-2 processing to its mature form, while treatment with the statin drug mevinolin to deplete sterols activates SREBP-2 processing (Western blot in bottom panel, the nuclear protein Lamin B1 serves as a loading control). In these conditions, mRNA levels of STARD4 and HMG CoA reductase are coordinately regulated (quantitative RT-PCR in upper panel). **b** STARD4 protein levels show the same pattern of regulation by Western blot (upper panel, cytosolic beta-actin serves as a loading control) with quantification (lower panel). Adapted from Rodriguez-Agudo et al. [12]. *SREBP* sterol regulatory element binding protein, *mRNA* messenger ribonucleic acid, *RT-PCR* reverse transcription polymerase chain reaction

lation of STARD4 mRNA was confirmed in 3T3-L1 fibroblasts and human THP-1 macrophages, and extended to the protein level by STARD4 Western blots, even correlating with the amount of mature SREBP-2 (Fig. 7.2) [12]. STARD4 mRNA is also regulated with HMGCR in other cells culture models where SREBP is affected: cholesterol-loaded macrophages [13], THP-1 cells differentiated into macrophages [12], mouse embryonic fibroblasts deficient in a sterol dehydrogenase involved in cholesterol synthesis [14], and HepG2 cells cultured in lipoprotein-depleted serum [15]. Notably, hepatic STARD4 mRNA and protein levels are induced threefold by statin treatment of mice [16]. Second, transgenic and knockout mice with altered SREBP activities show the expected changes in STARD4 levels: mice transgenic for constitutively active nuclear SREBP-1a or SREBP-2 have higher STARD4 expression, whereas SCAP knockout mice with loss of SREBP activity have decreased STARD4 expression [17]. Furthermore, in the same nuclear SREBP transgenic mice, STARD4 is more highly regulated by SREBP-2 than SREBP-1 [13, 17], supporting a role in sterol metabolism rather than fatty acid metabolism. Third, the promoter for STARD4 in mice and humans has several potential SREs, and a proximal promoter fragment confers sterol regulation in luciferase reporter assays. Sterol regulation of mouse and human STARD4 promoter reporters was abrogated when one potential SRE (called SRE-B) was mutated, or when either of the two nearby CCAAT box elements were mutated (presumably affecting binding of NF-Y, which cooperates with SREBPs) [13]. Fourth, direct binding of SREBP-2 to the STARD4 promoter was observed by chromatin immunoprecipitation followed by deep sequencing (ChIP-seq) [18]. This ChIP-seq experiment revealed 1,800 binding sites in mouse liver treated with lovastatin plus ezetimibe to maximize SREBP-2 processing, and the majority of these sites had an SRE consensus motif of AA(G/A) ATGGC. In retrospect, the functional STARD4 promoter SRE-B motif defined in reporter assays [13] has a sequence of AGGATGGA in mouse or TAAATGGA in human, each with only two disagreements (underlined) from the ChIP-seq consensus motif. Given this preponderance of evidence, there can be little doubt that STARD4 is a direct transcriptional target of SREBP-2. There are only two inconsistent cases in the literature, and both can be explained by the duration of time: (1) sterol depletion and overload of U2OS cells had the expected effects on SREBP-2 processing but failed to affect STARD4 protein levels as detected by Western blot [19] and (2) a 0.5 % cholesterol feeding experiment for one week failed to affect hepatic STARD4 protein levels [16]. In both cases, the treatment time, 2 h for U2OS cells and 1 week for cholesterol feeding, was likely too short to see changes in protein level.

STARD4 expression can be induced during steroidogenesis like StAR/STARD1. In MA-10 mouse Leydig tumor cells treated with cyclic adenosine monophosphate (cAMP), StAR mRNA is induced ~50-fold while STARD4 mRNA is induced only ~3-fold, and the similar small induction of HMGCR suggests that steroidogenesis may deplete cellular cholesterol and activate SREBP-2 [13]. A more recent experiment in the same MA-10 cell line showed ~20-fold cAMP induction of STARD4 protein by Western blot, and these authors propose a role for STARD4 in steroidogenesis [20]. Until further studies are performed, it remains unclear whether STARD4 is specifically regulated by steroidogenic stimuli.

One report indicates that STARD4 is regulated in the early phase of the ER stress response (Fig. 7.1c) [21]. Tunicamycin treatment of HeLa cells followed by cDNA subtraction identified STARD4 along with four well-known ER stress-induced genes; binding immunoglobulin protein (BiP), GRP94, CHOP, and Herp. STARD4 induction peaked at ~2.5-fold by 4–6 h of treatment, then came down by 12–24 h. This time course, as well as the different species and cell type, may explain why tunicamycin failed to induce STARD4 after 18–20 h in mouse NIH-3T3 cells [13]. A luciferase reporter driven by the human STARD4 promoter was activated by tunicamycin and other ER stressors thapsigargin, dithiothreitol (DTT), and brefeldin A, as well as by overexpression of the ER stress induced transcription factor activating transcription factor 6 (ATF6) [21]. Notably, overexpression of a dominant negative ATF6 prevented STARD4 reporter activation by thapsigargin. Site-directed mutagenesis of three potential ER-stress response elements (ESREs CCAAT-N9-CCACG) individually in reporters suggested that the second element was most responsible for ER-stress response. Compared to the other human STARD4 luciferase reporters [13], this functional ESRE-like element lies between potential SRE-B and SRE-C. Other SREBP-2 target genes were not studied in this system, and it is notable that in mouse 3T3-L1 cells the SREBP-2 target, HMGCR, shows a similar time course of induction only in early ER stress [22]. It has also been observed that STARD4 expression increases during the differentiation of THP-1 cells from monocytes to macrophages, even at days 5–6 when mature SREBP-2 decreases, but there are increased levels of mature ATF6 [12]. Given that SREBPs and ATF6 are processed by the same proteases after ER to Golgi translocation [23] and may bind to the same DNA elements and coordinately regulate target genes [24]—and that STARD5 is induced by ER stress (see below)—there is potential cross-talk between cholesterol metabolism and ER stress such that regulation of STARD4 by ER stress via ATF6 deserves further study.

Other regulatory effects on STARD4 expression have been reported. STARD4 mRNA is induced in granulosa cells of women with diminished ovarian reserve, an effect also observed for STARD1 [25], though the mechanism for this is uncertain. One report proposes that STARD4 is a target of the p53 transcription factor, based on a p53-binding region identified downstream of the gene by an early ChIP-seq method and p53-dependent gene regulation (though STARD4 expression was down rather than up) [26]. While some authors have hypothesized that this may relate to adipocyte biology [27], further reports connecting p53 and STARD4 are lacking. Taking a similar approach with the abundant genome-wide data now available in the Encyclopedia of DNA Elements (ENCODE) project [28], there are dozens of transcription factors in different cell types that can potentially bind to the STARD4 proximal promoter region in addition to the SREBPs as expected. Furthermore, STARD4 has at least three uncharacterized enhancer regions located in the 50 kb upstream, all of which have factor binding, DNaseI hypersensitivity, and histone marks.

Stard5/STARD5 Gene Regulation

STARD5 does not show regulation by the SREBP-related manipulations that affect STARD4 [13], yet it is induced by the ER stress response. Robust (up to 10-fold) increases in STARD5 mRNA have consistently been shown upon treatment with ER stressors such as tunicamycin, thapsigargin, brefeldin A, and dithiothreitol in mouse NIH-3T3 fibroblasts (Fig. 7.3) [13], human HK-2 kidney tubular cells [29], and mouse 3T3-L1 preadipocytes [22]. In the last study, a dose dependent protein induction by thapsigargin also was shown (using a commercial antibody from Santa Cruz, see below). The three main pathways of the ER stress response are mediated by the transcription factors ATF4, nuclear ATF6, and spliced Xbp1 (Fig. 7.1c), and only expression of spliced Xbp1—but not active ATF6 or ATF4—activated STARD5 expression in 3T3-L1 cells [22]. The mechanism for activation of STARD5 by ER stress and Xbp1 apparently lies outside the proximal promoter, as luciferase reporters driven by ~2000 bp upstream of human STARD5 or ~400 bp upstream of mouse STARD5 failed to show ER stress regulation [22, 13]. Instead, markedly more stable STARD5 mRNA was observed upon thapsigargin treatment of 3T3L1 cells in the presence of actinomycin D to inhibit new transcription, suggesting a

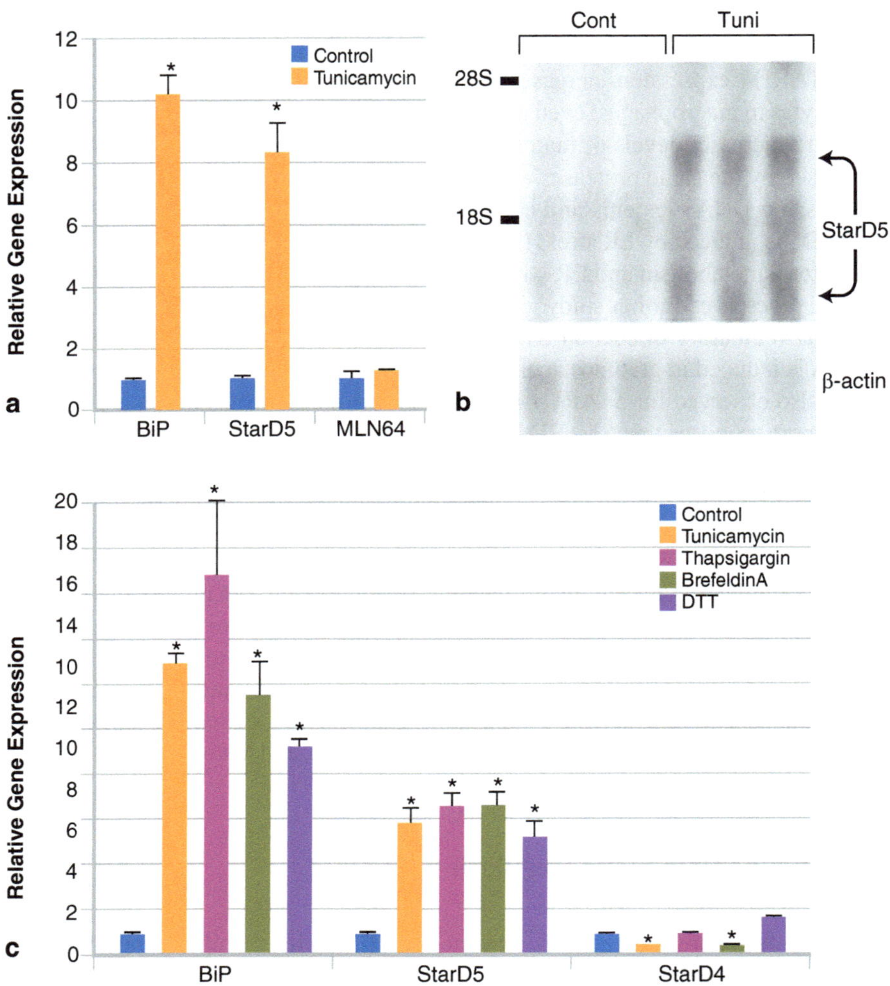

Fig. 7.3 STARD5 mRNA levels are increased in ER stress. In NIH-3T3 fibroblasts, treatment with drugs known to induce ER stress (tunicamycin, thapsigargin, brefeldin A, and DTT) all increases mRNA levels of STARD5 as well as the positive control ER stress response gene BiP. (**a**) and (**c**) show quantitative RT-PCR, while (**b**) shows a Northern blot with both STARD5 mRNAs upregulated. This research was originally published in reference 13, and subsequent studies in reference 22 also showed upregulation of STARD5 protein by Western blot, and that the mRNA induction is likely posttranscriptional via increased mRNA stability. Adapted from Soccio et al. [13] © the American Society for Biochemistry and Molecular Biology. *mRNA* messenger ribonucleic acid, *ER* endoplasmic reticulum

novel posttranscriptional mechanism to increase STARD5 mRNA levels [22]. Thus, even the STARD5 induction by spliced Xbp1 may be indirect via mRNA stability, though increased transcription remains possible and would require a nuclear run-on type assay to test. Some physiologic stimuli that induce ER stress also increased

STARD5 expression: cholesterol-loading of mouse macrophages [13] and OVE26 diabetic mouse kidney [29]. The exact role of STARD5 in the response to ER stress is a matter of speculation, as it is possible that lipids stressors as well as unfolded proteins play a role in ER stress. The ER membrane is notably poor in cholesterol, which constitutes only ~5% of its lipid molecules compared to ~30% in the plasma membrane (reviewed in ref. [19]).

STARD5 was regulated by the cytokine interleukin (IL)-1β in rat Sertoli cells, though the mRNA was induced and the protein decreased [30]. In contrast, STARD4 expression was not regulated by IL-1β, even though precursor and mature SREBP-1 levels decreased (SREBP-2 was not reported). There are no other reports of inflammatory cytokines regulating STARD5, and the significance of this finding is unknown.

STARD4 Expression Pattern and Subcellular Localization

When overexpressed in HeLa cells, GFP-tagged STARD4 gives a diffuse cytosolic and nuclear localization, in contrast to the vesicular pattern observed for full-length STARD1 (mitochondria) and MLN64 (endosomes) [31], and the same result is seen by immunofluorescence [32]. However, when overexpressed in human keratinocytes [33] or U2OS osteosarcoma cells [19], STARD4 staining is not seen in the nucleus and its cytosolic localization is more punctate, with a more intense perinuclear region.

The most detailed study of STARD4 localization used a polyclonal antibody generated in the laboratory of Dr W. M. Pandak, Virginia Commonwealth University [12]. This antibody was well validated, recognizing in Westerns a sterol-regulated band of the predicted size in human THP-1 macrophages, and a slightly larger band in mouse liver than human—consistent with the predicted 224 amino acid mouse protein versus 205 in humans. In human liver, STARD4 immunostaining was observed in hepatocytes, and it was even more intense in Kupffer cells but absent in endothelial cells. This was confirmed by Western Blots showing expression in both hepatocytes and nonparenchymal cells. In mouse 3T3-L1 cells, basal STARD4 immunostaining was weak and homogenous with some punctate regions, whereas sterol-depletion to increase expression resulted in stronger staining with a peri-nuclear reticular pattern, colocalizing with the ER marker calnexin. Fractionation of these cells confirmed cytoplasm and membrane association, without any detectable protein in nuclei or mitochondria. In THP-1 macrophages, STARD4 similarly co-localized with the ER marker calnexin, but also notably around ER-derived vesicles with bodipy-stained neutral lipids where acyl-coenzyme A cholesterol acyltransferase (ACAT)1 co-localized extensively (Fig. 7.4). Therefore, STARD4 appears to be a widely expressed cytosolic protein that may associate with ER membranes.

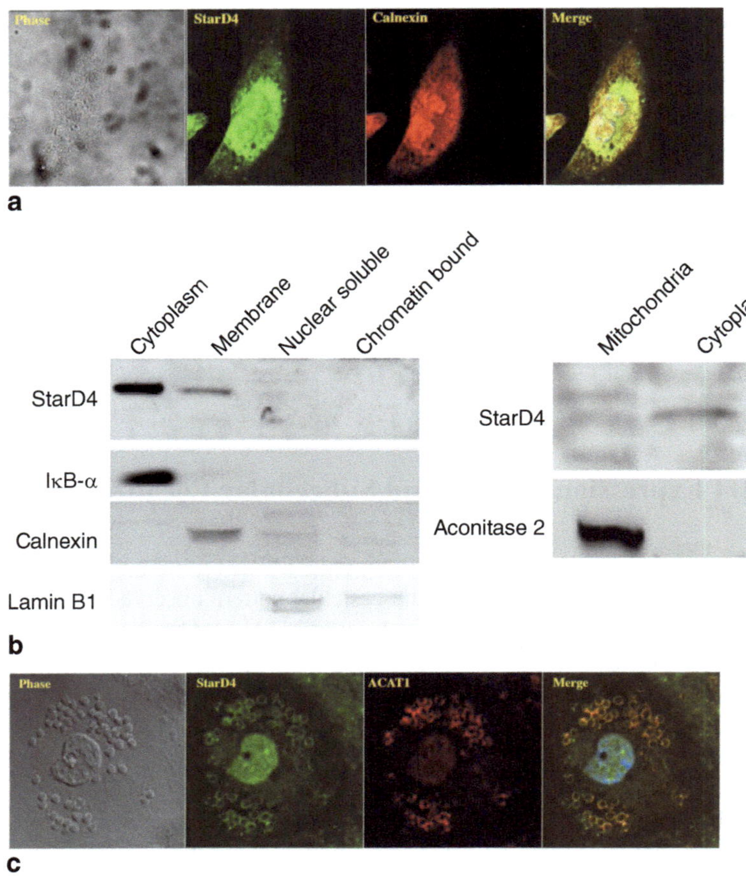

Fig. 7.4 STARD4 localization in 3T3-L1 cells and THP-1 macrophages. **a** A mouse 3T3-L1 fibroblast visualized by phase contrast microscopy (*left*) and stained with STARD4 (*green*), calnexin (ER maker, *red*), and DAPI (nucleus, *blue*). STARD4 is present in the cytosol colocalizing extensively with calnexin. **b** STARD4 Western blots were performed on 3T3-L1 cells fractionated into cytosol (marked by IκBα), membranes (marked by Calnexin), nuclei (marked by Lamin B1), and mitochondria (marked by aconitase 2), showing a predominantly cytosolic distribution with some membrane association. **c** A human THP-1 macrophage, in which phase contrast microscopy (left panel) show the nucleus and lipid droplets. Staining was performed for STARD4 (*green*), ACAT1 (*red*), and nucleus 4',6-diamidino-2-phenylindol (DAPI, *blue*). STARD4 appears present in the nucleus and the cytosol, where it is enriched surrounding lipid droplets and co-localizing with ACAT1. Adapted from Rodriguez-Agudo et al. [12]. *ER* endoplasmic reticulum

STARD5 Expression Pattern and Subcellular Localization

RNA expression of STARD5 is detected across multiple mouse tissues, with highest levels in liver and kidney [7]. A polyclonal antibody was generated in the Pandak lab against recombinant human STARD5 and detects a ~28kD band in human liver lysates, with subcellular fractionation showing this band in the cytosolic fraction but not the mitochondria or microsomes [34]. However, separation of liver cell types surprisingly showed STARD5 was not present in the isolated hepatocytes, but rather in the nonparencymal fraction [35]. STARD5 immunostaining of liver identifies cells lining sinusoids that stain for CD68, or Kupffer cells. Consistent with this, STARD5 protein was not detected using this antibody in primary human hepatocytes, HepG2 cells, or HUVEC endothelial cells, but it is found in immune cell lines derived from monocyte/macrophages, promyelocytic cells, mast cells, and basophils. Protein expression by Western Blot was also not detected in brain tissue or cell lines from fibroblasts, osteosarcoma, or astrocytes. For all of these Western blot analyses, the corresponding mRNA expression is not published, so it remains uncertain whether mRNA and protein expression correlate. For instance, multiple tissue northern blots show minimal STARD5 mRNA in spleen [7, 36] where many immune cells reside. Furthermore, a different polyclonal antibody against STARD5 generated in the Breslow lab (The Rockefeller University) recognizes a band in HepG2 hepatocellular carcinoma cells [15]. Unfortunately, no publications compare these two antibodies side-by-side, or validate either antibody by showing ER stress regulation or loss of signal in STARD5 knockdown. Further studies will be necessary to definitively address the cell type selective expression of STARD5 protein.

In THP-1 macrophages, STARD5 immunofluorescence with the Pandak laboratory antibody was located in the cytosol with a focal intensity near the nucleus co-localizing with the GM130 Golgi marker and high levels of free cholesterol by filipin staining [35]. Consistent with Golgi localization, STARD5 staining dispersed along with GM130 when the Golgi was disrupted with nocodazole, and it did not colocalize with a marker of the endocytic recycling compartment (ERC). However, fractionation and Western blotting for STARD5 showed only cytosolic localization and none in Golgi or ER, suggesting the observed Golgi association may be loose [35].

The Breslow laboratory antibody was used to study STARD5 expression and localization in kidney, recognizing a ~22kD band in mouse bone marrow derived macrophages and kidney [29]. Immunohistochemistry of mouse kidney shows STARD5 localization to proximal tubules in the cortico-medullary region and transitional epithelium lining the renal pelvis, but not in glomeruli. Immunoelectron microscopy showed STARD5 in cytosol and the apical membrane brush border at the base of microvilli where endocytosis occurs. STARD5 also associated with ER but not Golgi or mitochondria in renal tubule cells. In human HK-2 kidney tubule cells, STARD5 colocalized with the ER marker Grp78, both before and after treatment with the ER stressor tunicamycin, which relocated both markers from a diffuse punctate pattern to perinuclear and peripheral pattern. Also, STARD5 did not co-localize with an endosomal marker. The authors of this report suggest that differences in subcellular localization in renal tubule (ER and apical membrane) and macrophages (Golgi, see above) may reflect association with cholesterol-rich membranes rather than with a specific compartment [29]. As noted above and discussed further below, different antibodies could also be a factor.

Nuclear localization was not observed using the Pandak laboratory antibody for endogenous STARD5 in THP-1 macrophages, or primary hepatocytes overexpressing recombinant adenoviral STARD5 [35]. Likewise, using the Breslow laboratory antibody, STARD5 was located in the cytosol but not the nucleus of rat Sertoli cells [30]. However, in mouse 3T3-L1 cells assayed by immunofluorescence and subcellular fractionation using a commercial Santa Cruz antibody, STARD5 was located primarily in the nucleus, though remarkably, it redistributed to the cytosol and perinuclear locations after ER stress with thapsigargin [22]. This regulated subcellular localization of STARD5 is extremely interesting if validated in other systems.

A Note on STARD4 and STARD5 Antibodies

As detailed above, multiple polyclonal antibodies generated in different laboratories have been used to study STARD4 and STARD5, often with conflicting results, particularly for STARD5. Fortunately, there are now commercial antibodies available from Santa Cruz that seem to recognize proteins with the expected regulatory patterns. A STARD5 antibody from Santa Cruz gives a band in 3T3-L1 cells with the expected regulation by ER stress [22], and indeed this antibody showed the regulated nuclear localization above but was not used in prior localization or expression studies. A STARD4 antibody from Santa Cruz (sc-66663) shows loss of the STARD4 band by Western Blot in livers of knockout mice, and decrease in heterozygous mice [16]. The standard use of such well-validated commercial antibodies for STARD4 and STARD5 in the future may help clarify ambiguous and conflicting results about cellular and subcellular localization.

Structural Studies of STARD4 and STARD5

X-ray crystal structures have been published for STARD4 [37] and STARD5 [38], as have nuclear magnetic reasonance (NMR) solution structure models [39, 40]. STARD4 and STARD5 are very similar to each other and to other START domains: globular with a helix-grip fold, such that the curved β-sheet and C-terminal α-helix enclose a hydrophobic cavity large enough to accommodate a single lipid molecule. The four crystal structures of presumed cholesterol-binding START domains (StAR, MLN64, STARD4, and STARD5) all lack any lipid in the cavity, even when efforts were made to include cholesterol in the crystallization solutions. Models for reversible cholesterol binding have been proposed (reviewed in ref [41] and elsewhere in this volume). Comparative structural analysis has identified cavity residues that may mediate ligand specificity [38], and volumetric modelling of the hydrophobic cavities correlates with lipid binding [42]. While structure-function correlations have been performed extensively for STARD1, particularly mutations that cause lipoid congenital adrenal hyperplasia, similar studies have yet to be performed for STARD4 family members. It is obviously of great interest to determine which lipids occupy the binding cavities of STARD4 and STARD5.

STARD4 Lipid Binding and Transfer

STARD4 and STARD5 were initially predicted to bind cholesterol or related sterols based on sequence and structural similarity to STARD1 and STARD3 (MLN64) [7]. Consistent with this, STARD4 and STARD5 both show binding to fluorescently labelled NBD-cholesterol similar to that for STARD1 and STARD3 [43]. Given the volume of the lipid-binding cavity and the size of the nitrobenzoxadiazole (NBD) fluorophore, it is surprising that these START domains specifically bind to NBD-cholesterol at all, but lack of binding by STARD7 served as a negative control. Table 7.1 summarizes all the lipid-binding data reported to date, which is described below.

A direct-binding assay showed STARD4 binding to radiolabelled cholesterol but not 25-hydroxycholesterol or 27-hydroxycholesterol, and no competition by other unlabelled oxysterols [44]. Lipid protein overlay (LPO) assays were also performed, in which sterols were spotted on nitrocellulose membranes, then recombinant GST-STARD4 added and washed prior to detection with anti-Glutathione S-transferase (GST) antibody. By this assay, STARD4 was able to bind cholesterol strongly and 7α-hydroxycholesterol weakly (both were competed away by prebinding GST-STARD4 with cholesterol), but not other sterols. Circular dichroism (CD) spectroscopy also showed cholesterol causes dose-dependent changes in STARD4 far ultraviolet (UV) spectra, consistent with a conformational change upon binding [44].

In vitro lipid transfer assays test for the ability of a purified protein to facilitate movement of lipid from a donor to an acceptor membrane. Like StAR/STARD1, STARD4 was able to increase by 2.6-fold the transfer of [^{14}C]cholesterol from small unilamellar liposomes (50 nm small, unilamellar vesicles, SUVs) to isolated mitochondria [20]. Notably, STARD4 failed to transport normal or peroxidated phosphitidylcholine. STARD4 was even more effective at increasing the transfer of 7α-hydroperoxycholesterol, which the authors go on to show causes oxidative damage to mitochondria and loss of membrane potential, suggesting a potential deleterious effect of START domain mediated lipid transfer of oxidated lipids [20].

Another elegant transfer assay used donor liposome containing the florescent cholesterol analog dihydroergosterol (DHE) and acceptor liposomes with a florescent phopholipid, such that DHE transfer from donor to acceptor results in fluorescence resonance energy transfer (FRET) between the two lipids [19]. In this system, STARD4 was an extremely efficient transporter: it increased the rate of DHE transfer by 5 orders of magnitude versus the spontaneous level, with 1 uM STARD4 as effective as 1000 uM cyclodextrin, and each molecule of STARD4 transferred 7 DHE molecules per minute. STARD4 has a positively charged patch on its surface that was hypothesized to interact with negatively charged lipid head groups. Indeed, the presence of anionic lipids phosphatidylserine and phosphatidylinositol in donor and acceptor membranes increased STARD4 transfer activity 10-fold relative to neutral liposomes. Mutation of lysines in the STARD4 basic patch decreased trans-

Table 7.1 Lipid binding and transfer assays for STARD4 and STARD5. Various assays have been used to measure lipid binding by STARD4 and STARD5, sometimes with STARD1 as a control: direct in vitro binding of labelled lipid to recombinant tagged protein (Binding), competitive binding of unlabelled lipid to displace a labelled lipid (Competition), nuclear magnetic reasonance (NMR), lipid protein overlay (LPO), and transfer of lipid between membranes (Transfer). Relative binding strength is indicated by (+) signs, no binding by (−), no binding when another assay was positive by (−*), and assay not reported by (N/A). See text for details

Reference	Lipid	Assay	STARD1	STARD4	STARD5
43	*NBD-cholesterol*	Binding	+++	+++	+++
34, 44	*Cholesterol*	Binding	+++	+++	+++
	25-hydroxycholesterol	Binding	−	−	+
	27-hydroxycholesterol	Binding	N/A	−	−
	Cholesterol	Competition	N/A	N/A	+++
	24-hydroxycholesterol	Competition	N/A	−	−
	25-hydroxycholesterol	Competition	N/A	N/A	−*
	24, 25-hydroxycholesterol	Competition	N/A	−	−
	7α-hydroxycholesterol	Competition	N/A	−*	−
	Cholic acid	Competition	N/A	N/A	−*
	β-sitosterol	Competition	N/A	N/A	−
45–47	Cholesterol	NMR	+++	N/A	−
	25-hydroxycholesterol	NMR	N/A	N/A	−*
	Deoxycholic acid	NMR	N/A	N/A	+++
	Lithocholic acid	NMR	N/A	N/A	+++
	Chenodeoxycholic acid	NMR	N/A	N/A	++
	Glycodeoxycholic acid	NMR	N/A	N/A	+
	Taurodeoxycholic acid	NMR	N/A	N/A	+
	Cholic acid	NMR	N/A	N/A	+
	Ursodexoycholic acid	NMR	N/A	N/A	+
	Taurocholic acid	NMR	N/A	N/A	−
	Glycocholic acid	NMR	N/A	N/A	−
44	*Cholesterol*	LPO	N/A	+++	N/A
	7α-hydroxycholesterol	LPO	N/A	+	N/A
	25-hydroxycholesterol	LPO	N/A	−	N/A
	27-hydroxycholesterol	LPO	N/A	−	N/A
	24, 25-hydroxycholesterol	LPO	N/A	−	N/A
	20α-hydroxycholesterol	LPO	N/A	−	N/A
	7-ketocholesterol	LPO	N/A	−	N/A
	Stigmasterol	LPO	N/A	−	N/A
20	*Cholesterol*	Transfer	N/A	++	N/A
	7α-hydroperoxycholesterol	Transfer	N/A	+++	N/A
	Phosphatidylcholine	Transfer	N/A	−	N/A
	Peroxidized phosphatidylcholine	Transfer	N/A	−	N/A
19	*DHE (cholesterol surrogate)*	Transfer	N/A	+++	N/A

fer activity with anionic liposomes by eightfold. Increasing unsaturated fatty acid chains in acceptor (but not donor) liposomal phosphatidylcholine also increased transfer activity 3-fold, and this is relevant because the ER is enriched for unsaturated acyl chains and STARD4 may favor transfer to this membrane (see below) [19]. Overall, STARD4 consistently binds and transfers cholesterol, and perhaps also some 7α-metabolites.

STARD5 Lipid Binding

In vitro binding assays have shown that STARD5 can bind several sterol and non-sterol ligands. One direct binding assay uses recombinant bacterially generated His-tagged human START domains to bind radiolabelled sterols, retaining binding through washes of the nickel-resin and elution. STARD5 was able to bind both [^{14}C]cholesterol and [^{3}H]25-hydroxycholesterol, though cholesterol with higher affinity since in competitive assays it could displace 25-hydroxycholesterol but not vice versa [34]. Cholesterol binding was similar to StAR/STARD1, with saturable 1:1 stoichiometry; however, STARD1 did not bind 25-hydroxycholesterol. Furthermore, STARD5 did not bind [^{14}C]27-hydroxycholesterol, and no competition with [^{14}C]cholesterol was observed using other unlabelled sterols or cholic acid, though this result is difficult to interpret since even 25-hydrocholesterol did not compete. CD spectroscopy showed cholesterol and 25-hydroxycholesterol cause dose-dependent changes in STARD5 far UV spectra, consistent with conformation change upon binding [34].

NMR spectroscopy is another approach taken recently to characterize STARD5 lipid binding (see Chap. 3 of this volume). Similar methods had shown cholesterol binding to STAR and STARD6, yet the same authors showed that STARD5 did not bind cholesterol or 25-hydroxycholesterol, but rather the bile acids cholic acid and chenodeoxycholic acid [45, 46]. The NMR structure of STARD5 was identical to the X-ray crystal structure, allowing identification of residues in contact with bile acids using the "SAR by NMR" (structure activity relationship) method. Remarkably, the residues all lined the internal lipid binding cavity of STARD5, demonstrating for the first time that a lipid could occupy this cavity. This also presumably rules out long-range allosteric effects or specific binding outside the cavity [46]. In these NMR studies, there was also little perturbation of the overall structure by lipid binding, in contrast to the change in CD spectra above. These authors went on to report the relative affinity of STARD5 for multiple common bile acids, showing STARD5 has the highest affinity—in the physiological range—for unconjugated secondary bile acids without 7α-OH groups (deoxycholic and lithocholic acid) [47]. Overall for STARD5, there is a clear discrepancy between biochemical studies showing cholesterol (and 25-hydroxycholesterol) binding and biophysical NMR studies showing bile acid binding but not cholesterol—even though other START domains of STARD1 and STARD6 bind cholesterol as demonstrated using NMR technology.

STARD4 and STARD5 Activities

There are several pathways in intracellular cholesterol transport that are thought to be non-vesicular and mediated by intracellular transport proteins, including transport to mitochondria mediated by StAR/STARD1 in steroidogenic tissues (reviewed in ref. [48]). Table 7.2 summarizes START domain activities in steroidogenesis and

Table 7.2 Activities observed upon overexpression of STARD4 and STARD5. Various functional assays have been performed in cells overexpressing steroidogenic acute regulatory protein (StAR)-related lipid transfer (START) domain proteins. Relative activities are indicated by (+) signs, no activity by (−), and assay not reported by (N/A). Some activities (multiple) are shared among the presumed cholesterol-binding domains of STARD1, STARD3, STARD4, STARD5, and STARD6 but not the phospholipid-binding STARD2/PCTP. Other activities were selective for STARD4 or STARD5 and not seen for other proteins. See text for details

	Activity	Reference	STARD1	STARD3	STARD4	STARD5	STARD6	STARD2/PCTP
Multiple	Increase steroidogenesis (cell culture)	13	++++	+++	++	++	N/A	−
	Increase steroidogenesis (isolated mitochondria)	43	+++	+++	+	−	++++	N/A
	Increase bile acid synthesis (via mitochondrial Cyp27)	44	++++	N/A	+++	−	N/A	N/A
	Activate LXR reporter	2	++++	++++	+++	++	N/A	−
	Repress SREBP reporter	unpub	+++	+++	++	++	N/A	−
STARD4 selective	Increase cholesterol ester levels	44	−	N/A	++	−	N/A	N/A
	Increase in vitro ACAT activity	12	N/A	N/A	++	−	N/A	N/A
STARD5 selective	Increase filipin staining free cholesterol	44	−	N/A	−	+++	N/A	N/A
	Activate mRNA expression of SREBP-2 and LXRα	51	N/A	−	−	++	N/A	N/A

other functional assays, described below. In one of the most common steroidogenesis assays, non-steroidogenic COS-1 cells are transfected with the mitochondrial P450 side change cleavage enzyme system and 3β-hydroxysteroid dehydrogenase. These two enzymes convert cholesterol to pregnenolone then progesterone, which is drastically increased by StAR-like activity of delivering cholesterol to mitochondria. In this cell culture transfection assay, STARD4 and STARD5 increased progesterone production two–threefold, less than the five–sevenfold for STARD1 and STARD3/MLN64 START domains (Fig. 7.5a) [13]. When these same START domains were localized to the cytosolic side of the outer mitochondrial membrane by fusion with the Tom20 protein (as in ref. [49]), then all four had equal steroidogenic activity (Fig. 7.5b, unpublished data from Soccio and Breslow). However, when isolated mitochondria were used for similar steroidogenesis assays, high activity was observed for STARD1, STARD3, and STARD6, whereas STARD4 had minimal activity and STARD5 had none [43]. When overexpressed by adenovirus in primary mouse hepatocytes, both STARD1 and STARD4 could increase the rate of bile acid synthesis, but STARD5 lacked this activity [44]. Since these cells only use the alternative/acidic pathway of bile acid synthesis initiated by the mitochondrial Cyp27 enzyme, this result is also interpreted as cholesterol delivery to mitochondria. Together, these data suggest that STARD4 and STARD5 may be capable of delivering cholesterol to mitochondria under some conditions, but with less activity than STARD1—even N-62 STAR lacking the mitochondrial import signal.

STARD4 and STARD5 may have roles in cholesterol synthesis and esterification, both of which involve the ER, and thus affect cellular levels of free and esterified cholesterol (see Chap. 8 of this volume for further detail). STARD4 overexpression in primary mouse hepatocytes increases neutral lipid staining by Oil Red O and cholesterol ester production from [^{14}C]cholesterol, effects not observed for STARD1 or STARD5 [44]. Remarkably, even isolated mouse liver microsomes showed increased ACAT activity (incorportation of [^{14}C]oleoyl-CoA into cholesterol ester) in the presence of STARD4 but not STARD5 [12]. Notably, overexpressed STARD4 did not affect the production of cholesterol or cholesterol esters from [^{14}C]acetate, showing that STARD4 does not increase cholesterol biosynthesis or the esterification of newly synthesized cholesterol [44]. Together, these data indicate that STARD4 increases esterification of preformed but not newly synthesized cholesterol—suggesting the existence of distinct pools of subcellular cholesterol.

STARD4 knockdown was reported in HepG2 human hepatoma cells by stable short hairpin RNA (shRNA) and resulted in an ~50% decrease in the RNA and protein levels under conditions of high expression in lipoprotein-depleted serum [15]. Consistent with STARD4 delivery of cholesterol to ACAT, STARD4 knockdown cells had ~40% decreased cholesterol ester levels and ~60% decreased ACAT activity (measured by incorporation of [^{14}C]oleate into cholesterol esters). While free cholesterol levels overall were unchanged, filipin staining of free cholesterol showed more cholesterol at the plasma membrane in STARD4 knockdown cells, both in cholesterol-depleted and -replete conditions. Along with apparently more cholesterol in the plasma membrane, there was ~70% less free cholesterol in ER membrane fractions from STARD4 knockdown cells. Furthermore, cholesterol

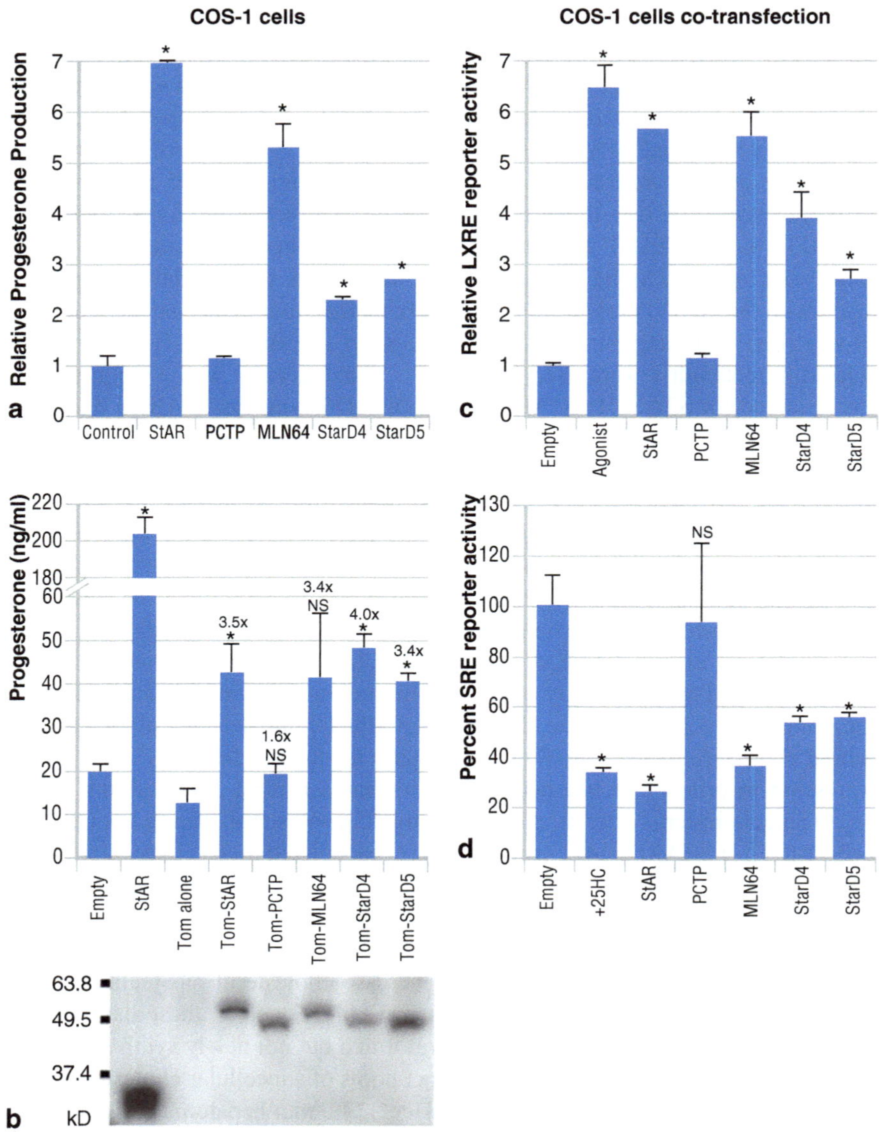

Fig. 7.5 Some activities common to STARD4, STARD5, STARD1/STAR, and STARD3/MLN64. **a** A co-transfection steroidogenesis assay was performed in COS-1 cells (see text for details), and the four cholesterol-binding START domains (but not STARD2/PCTP) were able to stimulate progesterone production. STARD1 and STARD3/MLN64 consistently showed higher activity than STARD4 and STARD5. **b** The same steroidogensis assay was performed with Tom20 fusion proteins localizing the START domains to the cytosolic face of outer mitochondrial membrane. In this case, the four cholesterol-binding START domains had similar activities, with the anti-FLAG tag Western blot in the lower panel showing similar expression levels. **c** COS-1 cells were transfected with expression plasmids and luciferase reporters for LXR activity, with a pharmacologic LXR agonist serving as the positive control. The four cholesterol-binding START domains (but not PCTP) were able to activate the LXR reporter to various degrees in the absence of agonist. **d** HEK-293 cells were transfected with an SRE-luciferase reporters to measure endogenous SREBP activity,

Table 7.3 STARD4 knockdown and overexpression give opposite effects. STARD4 knockdown was reported in HepG2 cells (reference 15) and overexpression in U2OS cells (reference [19]), giving consistent opposite effects in various functional assays of intracellular cholesterol transport and metabolism. These results support a role of STARD4 in delivering cholesterol from the plasma membrane to intracellular membranes like the endocytic recycling compartment and the endoplasmic reticulum, where activities of affect acyl-coenzyme A cholesterol acyltransferase (ACAT) and sterol regulatory element binding protein (SREBP) cleavage-activating protein (SCAP) are affected. (See text for details.)

	Knockdown (HepG2, ref 15)	Overexpression (U2OS, ref 19)
ACAT activity	Decreased	Increased (DHE localization to lipid droplets)
Cholesterol ester levels	Decreased	Increased
ER cholesterol levels	Decreased	Increased (SCAP relocation to ER upon cholesterol loading)
Plasma membrane cholesterol	Increased	Not reported
Cholesterol (DHE) transfer to ERC	Slowed	Accelerated

movement to the ERC was studied using DHE and fluorescence recovery after photobleaching (FRAP). In control cells in lipoprotein depleted serum, adding cholesterol resulted in more rapid DHE recovery ($t_{1/2}$ from 118 to 94 s), while this effect was abrogated in STARD4 knockdown cells ($t_{1/2}$ not significantly changed from 110 to 117 s). Transferrin recycling was unaffected, indicating that STARD4 selectively affects non-vesicular sterol trafficking to the ERC. STARD5 protein levels were not affected by STARD4 knockdown, but under cholesterol-depleted conditions STARD4 knockdown cells had increased cell surface LDL receptor and decreased NPC1 protein (an endosomal transmembrane cholesterol transporter), suggesting compensatory responses [15].

Elegant experiments involving STARD4 knockdown and overexpression were also reported in U2OS human osteosarcoma cells [19]. Table 7.3 summarizes the consistently opposite effects reported for STARD4 knockdown and overexpression. STARD4 overexpression had many notable effects: ACAT-dependent re-localization of DHE from the plasma membrane to neutral lipid droplets (i.e. esterification), increasing cellular cholesterol esters twofold, and increasing the rate and extent of DHE FRAP in the ERC. In cholesterol-depleted cells upon reloading with cholesterol, SCAP relocates from Golgi to ER resulting in less SREBP activation, and STARD4 accelerated relocation of SCAP—presumably by increasing ER cholesterol. Notably, microinjection of these cells with the nonselective lipid exchanger methylcyclodextrin (MCD) had all of the same effects delivering cholesterol to the ER and ERC. Conversely, small interfering RNA (siRNA) knockdown of STARD4

which is repressed by 25-hydroxycholesterol (25HC) as a positive control. The four cholesterol-binding START domains (but not PCTP) were able to repress SREBP activity. The research in (**a**) and (**c**) was adapted from Soccio [13]. © the American Society for Biochemistry and Molecular Biology. (**b**) and (**d**) are unpublished data from Soccio and Breslow. *MLN64* metastatic lymph node clone 64, *START* steroidogenic acute regulatory protein (StAR)-related lipid transfer, *SREBP* sterol regulatory element binding protein

in U2OS cells increased filipin staining without changing its expression pattern and increased free cholesterol levels by ~50 %, and these effects were rescued by expression of an siRNA-resistant STARD4 or by microinjection of MCD. This result can be interpreted based on ER cholesterol sensing by SCAP: lack of STARD4 results in less ER-free cholesterol, such that SREBP2 establishes a higher "set point" for cellular cholesterol. Furthermore, in the setting of cholesterol loading, STARD4 siRNA resulted in defective cholesterol esterification (measured by cholesterol ester content and transfer of DHE to lipid droplets), again rescued by MCD microinjection. Therefore, while STARD4 can apparently mediate transfer of cholesterol to the ER for esterification by ACAT or sensing by SCAP, the non-targeted carrier MCD can have the same effects.

The effects of STARD5 overexpression are different from STARD4. In the same mouse primary hepatocyte system where STARD4 affected cholesterol esterification, overexpression of STARD5 (but not STARD1 or STARD4) increased cellular free cholesterol 12-fold as assessed by filipin staining, similar to what was already observed in primary rat hepatocytes [34, 44]. A correlation between increased STARD5 and higher renal free cholesterol was also observed in the diabetic OVE26 mice [29], and the same positive correlation is seen in human proximal tubule cell lines [50]. However, STARD5 overexpression did not change the distribution of exogenously added DHE in U2OS cells, a system in which STARD4 caused DHE to localize to lipid droplets [19]. In THP-1 human macrophages, overexpression of STARD5 had drastic effects increasing mRNA expression of SREBP-2 and liver X receptor (LXR)α, whereas MLN64/STARD3 and STARD4 did not have these effects [51]. Therefore, while not affecting ACAT, STARD5 appears to affect cellular free cholesterol levels.

The SREBP and LXR transcription factors respond to cellular sterols, so perturbations in cellular sterol metabolism may affect their activities. Consistent with this, overexpression of STARD4 or STARD5 (as well as the START domains of STARD1 and STARD3) could activate luciferase reporters driven by the nuclear receptor LXR (Fig. 7.5c), and this result is interpreted as sterol transport to generate or translocate an LXR ligand oxysterol [13]. Likewise, overexpressed STARD4 and STARD5—again like the STARD1 and STARD3 START domains—could repress SRE driven-reporters, suggesting sterol transport affecting sterol-sensing by SCAP and thus SREBP activity (Fig. 7.5d, unpublished data from Soccio and Breslow). The effect of STARD4 on SREBP-2 processing was tested directly, and both overexpression and knockdown of STARD4 could affect SREBP-2 cleavage in complex ways [19]. Since SREBP-2 activates STARD4, and STARD4 can deliver cholesterol to the ER, then STARD4 negatively feeds back on SREBP activity. Statin drugs rely on SREBP-induced LDL receptor expression to lower serum cholesterol, so statin-induced STARD4 expression (shown in ref. 16) may attenuate this effect such that STARD4 inhibition would increase statin efficacy. If this model is correct, then it is remarkably analogous to Pcsk9, which is induced by SREBP2 but triggers degradation of the LDL receptor, such that Pcsk9 inhibition is desirable to lower serum cholesterol [9]. STARD4 and Pcsk9 may thus both serve as negative feedback "brakes" on the SREBP-2 and LDL receptor pathway (Fig. 7.6).

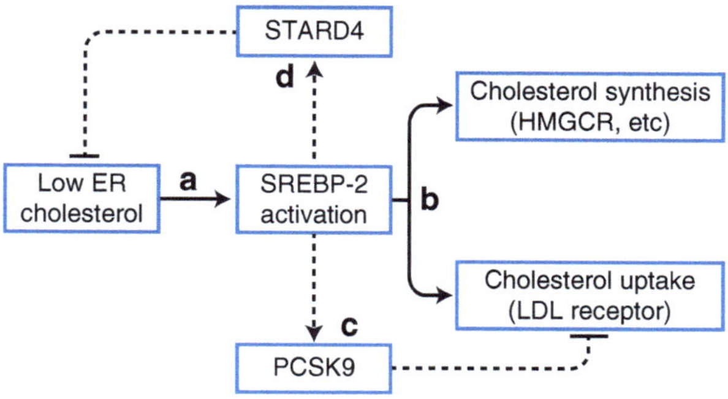

Fig. 7.6 Like Pcsk9, STARD4 may negatively feedback on the SREBP-2/LDL receptor pathway. **a** Low levels of ER cholesterol result in proteolytic activation of the SREBP-2 transcription factor. **b** Mature SREBP-2 activates transcription of genes involved in cholesterol synthesis (HMGCR and other biosynthetic enzymes) and uptake (the LDL receptor), thus restoring cholesterol levels. **c** SREBP-2 activates Pcsk9, which functions to inactivate the LDL receptor and thus blunt the effect of SREBP-2 on cholesterol uptake. **d** SREBP-2 also increases expression of STARD4, which can deliver cholesterol to the ER and thus decrease SREBP-2 activity. Widely used statin drugs rely on SREBP-2 mediated LDL receptor activation to lower serum LDL levels, and inhibition of Pcsk9 is known to potentiate this effect. Likewise, inhibition of STARD4 could increase the effects of statins on SREBP-2 and LDL receptor. *Pcsk9* proprotein convertase subtilisin/kexin type 9, *SREBP* sterol regulatory element binding protein, *LDL* low-density lipoprotein, *ER* endoplasmic reticulum, *HMGCR* HMG CoA (3-hydroxy-3-methylglutaryl coenzyme A) reductase

One small study with overexpression of STARD4 in human keratinocytes by transient transfection gave conflicting results, such as decreased incorporation of [^{14}C]acetate into free cholesterol and cholesterol esters, suggesting less cholesterol synthesis [33]. While no changes in "total lipid mass" were noted, it is unclear whether cholesterol esters were measured (the data above suggest STARD4 mediates increased esterification of preexisting cholesterol). There were also changes in gene expression for SREBP-2 (increased, without changes in target genes like HMGCR) and ABCA1 and ABCG4 lipid transporters (though these went in opposite directions despite both being LXR targets induced by oxysterols). These keratinocyte results are difficult to interpret, but support the idea that STARD4 can redistribute cellular sterols to affect lipid metabolic pathways.

In contrast to potential negative effects on statin efficacy, START protein-mediated lipid transfer may have beneficial effects on atherosclerotic disease (see Chap. 5 of this volume). Several studies have ectopically expressed STARD1 in non-steroidogenic cells and tissues that do not normally express high levels, such as hepatocytes, macrophages, and endothelial cells. STARD1 expression in THP-1 macrophages could reduce lipid accumulation and inflammation, likely by increasing LXR signaling and expression of target genes [52]. Similar effects on LXR target genes were seen in endothelial cells [53]. Likewise, viral infection to overexpress STARD1 predominantly in the liver of ApoE null mice resulted in lowering of serum cholesterol, as well as reduction in hepatic steatosis and—most notably—aortic atherosclerotic lesions (Fig. 7.7) [54]. It is likely that effects of STARD1

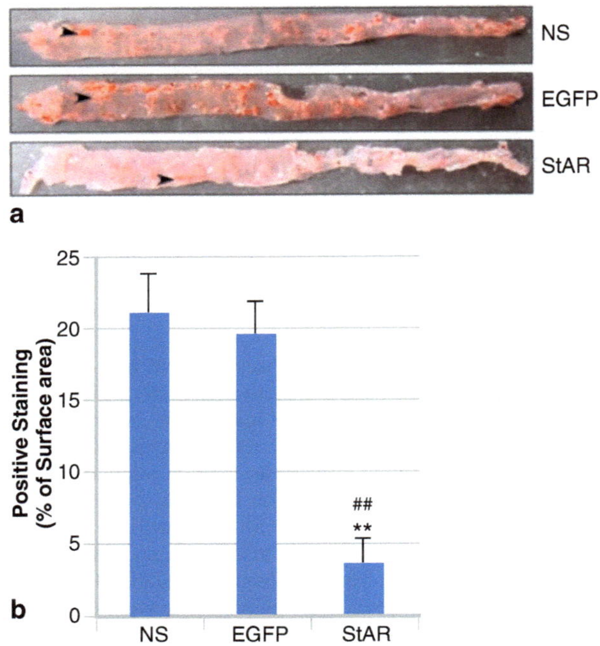

Fig. 7.7 Viral expression of STARD1/STAR decreases atherosclerosis. **a** ApoE null mice develop aortic atherosclerotic lesions, which stain with Sudan IV dye for neutral lipid in *en face* preparations. Mice infected with an adenovirus expressing STAR had markedly smaller lesions compared to control mice that were either uninfected (NS) or infected with a control virus (EGFP). **b** Quantification of lesion area confirms the protective effect of STAR. Notably, in this system STAR is predominantly overexpressed in the liver where it is not normally found, while STARD4 and STARD5 are normally expressed there. Adapted from Ning Y, Xu L, Ren S, Pandak WM, Chen S, Yin L. STAR overexpression decreases serum and tissue lipids in apolipoprotein E-deficient mice. Lipids. 2009 Jun;44(6):511–9. PMID: 19373502

in these ectopic experimental contexts may mimic the physiology of other related START proteins like STARD4 and STARD5 that are normally expressed in liver and macrophages. Overall, given their potential effects on subcellular cholesterol transport, SREBP processing, and LXR activation, loss of STARD4 or STARD5 protein activity may have complex phenotypes—and in the absence of pharmaceutical inhibitors, knockout mice are an excellent way to study these.

STARD4 Subfamily Knockout Mice

Given that STARD4 can clearly affect cellular cholesterol metabolism, it was hypothesized that the knockout mice might have phenotypes reflecting altered cholesterol transport. However, mice with a whole body null allele for STARD4 were born at normal Mendelian ratios, developed normally, and were apparently healthy with normal male and female fertility and normal litter sizes [16]. Pathological examination (gross and histological) as well as serum chemistries and blood counts failed to show any differences. On chow diet, the STARD4 knockout mice had lower body weights by approximately 2 g (males from weeks 5–12, though females differed significantly only at week 12), but they were also shorter in length such that the body mass index was unchanged. Food intake during weeks 6–8 did not differ significantly. When male mice were placed on a high fat diet for 12 weeks (weeks 8–20), there was no longer any significant weight difference. Dual-energy X-ray absorptiometry (DEXA) scans for body composition at week 8 (prior to high fat diet) and week 20 (after high fat diet) showed no difference in lean, fat, and bone mass. Notably, cholesterol levels in knockout mice did not differ in plasma (total, high-density lipoprotein (HDL), non-HDL, free, esterified) or liver (total, free, and esterified), nor did plasma or liver triglycerides. This lack of difference was seen in both sexes on chow diet, upon statin treatment, or upon cholesterol feeding (with the exception of some small decrease in serum total, LDL, and esterified cholesterol in female mice on high cholesterol diet). Fasting glucose and glucose tolerance were also unchanged. A subtle difference was reported in the bile, which appeared ~25 % more dilute in female knockout mice (based on biliary cholesterol, phospholipid, and bile acids, though the last component was not significantly different), but this difference was only seen in female but not male mice, and measurement was only reported on the chow diet, not the other diets tested. Notably, microarray profiling of liver gene expression on a cholesterol-free diet showed no significant differentially regulated genes in the knockout mice. Quantitative reverse transcription (RT)-PCR analysis of candidate genes showed some small differences that were not consistent across diets (i.e., NPC1 was decreased in knockout mice on a 0.0 % cholesterol diet, but not with lovastatin or 0.5 % cholesterol, while STARD5 was slightly increased in knockout mice on statin and 0.5 % cholesterol but not on 0.0 % cholesterol). In conclusion, any of the reported differences between wild-type and STARD4 knockout mice are subtle and inconsistent across conditions (gender, diet, etc.), suggesting that they could merely arise from chance variation when so many parameters were tested in so many combinations. This lack of any overall apparent phenotype suggests that STARD4 functions can be replaced by other redundant cellular proteins [16].

Despite the overall lack of a phenotype in STARD4 null mice, these animals nonetheless merit further study. For instance, stressing the mice with obesity and metabolic syndrome (such as high fat diet-fed or ob/ob mice, in which serum and hepatic lipids are markedly increased) or atherosclerosis (such as ApoE or LDL receptor null mice, in which macrophage cholesterol handling affects lesion progression) may unmask phenotypes. To address potential redundancy among START family members, combined knockout of STARD4 could be generated with STARD1, STARD3, or STARD5. Likewise, studies of isolated STARD4 null primary macrophages or hepatocytes may reveal effects on cholesterol uptake, esterification, or efflux that are compensated in the whole animal. Tissue specific knockouts could also be generated by using the existing LoxP-flanked STARD4 allele with different Cre recombinase drivers, and acute hepatic knockout by infection with adenoviral

Cre could reveal effects prior to compensatory mechanisms. Furthermore, studies of mice lacking STARD5 and STARD6 should be undertaken, particularly as these gene-targeted mice are now available via the International Knockout Mouse Consortium [55]. One obvious question is whether male mice lacking STARD6 are fertile, as some defect in sperm maturation or function may exist (see below). Likewise, an effect on bile acid homeostasis in STARD5 null mice would support the notion that STARD5 binds these cholesterol-derived molecules.

What About STARD6?

There have been many fewer studies of STARD6 compared to the other subfamily members STARD4 and STARD5. STARD6 expression by Northern Blot was observed only in the testis [7], and subsequent in situ hybridization studies in rats further localized the mRNA to male germ cells, particularly the maturing round and elongated spermatids [56]. An anti-STARD6 antibody confirmed testis-specific expression by Western Blot, and immunostaining in round and elongated spermatids [56]. This agrees with unpublished observations that STARD6 mRNA is neither detected in MA-10 or freshly isolated Leydig cells nor in the testis of Kit W/Wv mice [57] deficient in germ cells (Fig. 7.8, unpublished data from Soccio and Breslow). There is one conflicting report of STARD6 immunostaining of interstitial Leydig cells of hypothyroid but not control rats [58], but no mRNA or Western analysis was performed to confirm this unexpected localization in Leydig cells. STARD6 protein transfers cholesterol to mitochondria as efficiently as STARD1 [43], and notably the STARD6 gene falls in a mouse quantitative trail locus that affects activity of sperm mitochondrial diaphorase enzyme activity [59]. The localization of STARD6 in mature sperm has not been reported, and this will be informative as mitochondria localize to the midpiece. Cholesterol and its precursor sterols have several essential roles in sperm maturation and function (reviewed in ref. 7 and 56), and STARD6 knockout mice may have impaired fertility due to altered cholesterol metabolism in germ cells.

STARD6 expression in brain has reported by immunohistochemistry, in nuclei but not cytosol, of neurons and glial cells throughout the central and peripheral nervous system [60]. These authors used the same antibody described above, and could not explain the discrepancy that it failed to detect a band by Western blot in brain [56], nor that STARD6 mRNA is undetectable in brain [7]. Nonetheless, this group has gone on to publish a series of reports about STARD6 immunolocalization in various brain regions. STARD6 immunostaining in rat hippocampus was observed in the nucleus and cytoplasm and increased transiently after induction of seizures [61] or exitotoxicity [62], an effect actually confirmed by Western blots on hippocampal protein, and the localization was distinct from that of STARD1 immunolocalization [63]. STARD1 and STARD6 immunostaining were also reported in Purkinje cells of the cerebellum, potentially responding differentially to hypothyroidism [64]. A potential role in neurosteroid production is proposed, but this remains entirely speculative.

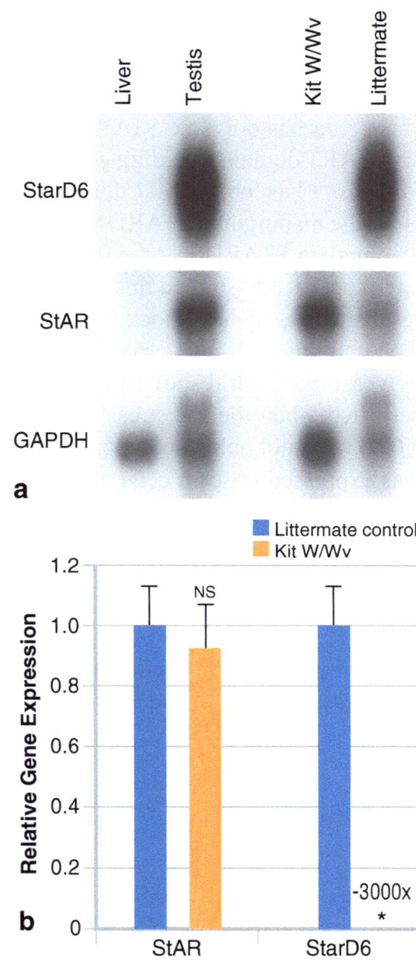

Fig. 7.8 STARD6 mRNA is absent in germ cell deficient testis from KitW/Wv mice. **a** Northern blots show abundant STARD6 and StAR mRNA in testis, but none in liver Glyceraldehyde 3-phosphate dehydrogenase (GAPDH) is an RNA loading control). In the testis of KitW/Wv mice which lack germ cells, STARD1 mRNA from Leydig cells remains present, while STARD6 mRNA is absent. The testis of littermate controls with germ cells show mRNA for both STARD6 and STARD1. **b** Quantitative RT-PCR analysis of STAR and STARD6 expression in testis of the same mice. Unpublished data from Soccio and Breslow. *StAR* steroidogenic acute regulatory protein, *mRNA* messenger ribonucleic acid, *RT-PCR* reverse transcription polymerase chain reaction

One report implicates STARD6 in non-small cell lung cancer cells response to the chemotherapeutic agent paclitaxel [65]. While this raises the interesting issue of tumors ectopically express spermatogenic proteins that may be functional in neoplasia, there are no other reports of STARD6 expression and function in tumors.

Regulatory elements mediating the male germ cell specific expression of STARD6 have not been identified. Notably, the STARD6 mRNA suggests many levels of regulation. 5′ RACE from testis cDNA identified multiple alternative first exons (three in mouse, two in human) as well as multiple initiation sites in these exons (unpublished data from Soccio and Breslow), showing alternative initiation and splicing. Also notable is the long multiexon 5′ UTR, which includes many potential ATG start codons which cannot encode the START protein, as a TAG stop codon immediately precedes the ATG for that open reading frame [7, 56]. Such upstream ATGs are rare in eukaryotic mRNAs and are thought to mediate translational regulation (reviewed in ref 56).

C. elegans Have One STARD4 Subfamily Member

While mammals have the three STARD4 subfamily members, the nematode worm *C. elegans* has only one STARD4 subfamily homolog, named K02D3.2, among its six START domain encoding genes [5]. (In comparison, the fruit fly *Drosophila melanogaster* has four START domain-encoding genes, and none of these is most similar to the mammalian STARD4 subfamily.) The K02D3.2 gene is ~23 % identical to mammalian STARD5, and only ~17 % to STARD4 and STARD6. No publications have specifically studied this gene, though two large screens have given suggestive results. First, siRNA knockdown of K02D3.2 resulted in decreased fat storage in the gut lining cells [66]. Second, a genome wide map of protein-protein interactions by yeast two-hydrid showed K02D3.2 binding to T01D1.6 (abu-11) [67], which is activated when the ER stress response is blocked [68]. The expression pattern of K02D3.2 is not published, though a GFP reporter driven by ~2.7 kb of its promoter was expressed in seam cells of the embryos and larvae, though it was not affected by cholesterol depletion or tunicamycin treatment (unpublished data, Soccio and Breslow in collaboration with Elliot Perens and Shai Shaham at the Rockefeller University). Seam cells actively secrete the cuticle covering the worm, so they may have high physiological levels of ER stress due to active protein secretion, but this is entirely speculative. Finally, K02D3.2 itself has not been implicated in the ER stress response in nematodes. Further studies specifically addressing K02D3.2 will be necessary to determine the role of this STARD4 subfamily protein in nematodes.

Summary

Despite a decade of work and dozens of publications, there are many unanswered questions regarding the STARD4 subfamily of START proteins. STARD6 expression in male germ cells clearly indicates a role in sperm and male fertility, while the consistent activation of STARD4 by SREBPs and STARD5 by ER stress suggests other potential functions. The preponderance of evidence generally indicates that all three proteins can bind and transfer cholesterol itself, while the results are inconsistent for related molecules (precursors, oxysterols, and bile acids). Given important roles for these other molecules in metabolism, ligand binding by these START proteins certainly merits further study, particularly comparative efforts looking at panels of START domains in the same assay system. STARD4 can clearly mediate intracellular cholesterol movement, particularly to ER-resident ACAT for esterification, though many of the cellular effects can be mimicked by non-selective transport by cyclodextrin. The STARD4 null mice have unchanged levels of cholesterol and cholesterol ester in liver and plasma, suggesting that other redundant sterol transfer mechanisms may exist. While not affecting esterification, STARD5 can have other drastic effects on cellular cholesterol metabolism that require further characterization, such as large increases in free cholesterol staining with filipin. STARD4 and

STARD5 appear to be widely expressed cytosolic proteins, though two intriguing aspects of STARD5 localization deserve careful study: the potential selective localization of STARD5 to macrophage and immune cells and the potential translocation from nucleus to cytoplasm upon ER stress.

In conclusion, there has been considerable progress in the understanding of the STARD4 subfamily's cellular and subcellular localization, regulation, lipid binding and transfer, and effects of knockdown and overexpression. This progress has yet to reveal their functions in normal and disease physiology, but many of the reagents and animal models to probe these questions now exist. There is a trend towards more publications naming STARD4 and/or STARD5 in their titles, as more than half of these have appeared since 2010 (10 of 19 total, with the other 9 from 2002–2009). The next decade of research on the STARD4 subfamily will hopefully show even more accelerated progress.

References

1. Lin D, Sugawara T, Strauss JF 3rd, Clark BJ, Stocco DM, Saenger P, Rogol A, Miller WL. Role of steroidogenic acute regulatory protein in adrenal and gonadal steroidogenesis. Science. 1995 March 24;267(5205):1828–31. PMID: 7892608.
2. Ponting CP, Aravind L. START: a lipid-binding domain in StAR, HD-ZIP and signalling proteins. Trends Biochem Sci. 1999 April;24(4):130–2. PMID: 10322415.
3. Tsujishita Y, Hurley JH. Structure and lipid transport mechanism of a StAR-related domain. Nat Struct Biol. 2000 May;7(5):408–14. PMID: 10802740.
4. Breslow JL. Human apolipoprotein molecular biology and genetic variation. Annu Rev Biochem. 1985;54:699–727. PMID: 3896129.
5. Soccio RE, Breslow JL. StAR-related lipid transfer (START) proteins: mediators of intracellular lipid metabolism. J Biol Chem. 2003 June 20;278(25):22183–6. PMID: 12724317.
6. Tiemann M, Han Z, Soccio R, Bollineni J, Shefer S, Sehayek E, Breslow JL. Cholesterol feeding of mice expressing cholesterol 7alpha-hydroxylase increases bile acid pool size despite decreased enzyme activity. Proc Natl Acad Sci U S A. 2004 Feb 17;101(7):1846–51. PMID: 14762172.
7. Soccio RE, Adams RM, Romanowski MJ, Sehayek E, Burley SK, Breslow JL. The cholesterol-regulated StarD4 gene encodes a StAR-related lipid transfer protein with two closely related homologues, StarD5 and StarD6. Proc Natl Acad Sci U S A. 2002 May 14;99(10):6943–8. PMID: 12011452.
8. Maxwell KN, Soccio RE, Duncan EM, Sehayek E, Breslow JL. Novel putative SREBP and LXR target genes identified by microarray analysis in liver of cholesterol-fed mice. J Lipid Res. 2003 Nov;44(11):2109–19. PMID: 12897189.
9. Petrides F, Shearston K, Chatelais M, Guilbaud F, Meilhac O, Lambert G. The promises of PCSK9 inhibition. Curr Opin Lipidol. 2013 Aug;24(4):307–12. PMID: 23817198.
10. Sahakitrungruang T, Soccio RE, Lang-Muritano M, Walker JM, Achermann JC, Miller WL. Clinical, genetic, and functional characterization of four patients carrying partial loss-of-function mutations in the steroidogenic acute regulatory protein (StAR). J Clin Endocrinol Metab. 2010 Jul;95(7):3352–9. PMID: 20444910.
11. Radhakrishnan A, Goldstein JL, McDonald JG, Brown MS. Switch-like control of SREBP-2 transport triggered by small changes in ER cholesterol: a delicate balance. Cell Metab. 2008 Dec;8(6):512–21. PMID: 19041766.

12. Rodriguez-Agudo D, Calderon-Dominguez M, Ren S, Marques D, Redford K, Medina-Torres MA, Hylemon P, Gil G, Pandak WM. Subcellular localization and regulation of StarD4 protein in macrophages and fibroblasts. Biochim Biophys Acta. 2011 Oct;1811(10):597–606. PMID: 21767660.
13. Soccio RE, Adams RM, Maxwell KN, Breslow JL. Differential gene regulation of StarD4 and StarD5 cholesterol transfer proteins. Activation of StarD4 by sterol regulatory element-binding protein-2 and StarD5 by endoplasmic reticulum stress. J Biol Chem. 2005 May 13;280(19):19410–8. PMID: 15760897.
14. Cunningham D, Swartzlander D, Liyanarachchi S, Davuluri RV, Herman GE. Changes in gene expression associated with loss of function of the NSDHL sterol dehydrogenase in mouse embryonic fibroblasts. J Lipid Res. 2005 June;46(6):1150–62. PMID: 15805545.
15. Garbarino J, Pan M, Chin HF, Lund FW, Maxfield FR, Breslow JL. STARD4 knockdown in HepG2 cells disrupts cholesterol trafficking associated with the plasma membrane, ER, and ERC. J Lipid Res. 2012 Dec;53(12):2716–25. PMID: 23033213.
16. Riegelhaupt JJ, Waase MP, Garbarino J, Cruz DE, Breslow JL. Targeted disruption of steroidogenic acute regulatory protein D4 leads to modest weight reduction and minor alterations in lipid metabolism. J Lipid Res. 2010 May;51(5):1134–43. PMID: 19965609.
17. Horton JD, Shah NA, Warrington JA, Anderson NN, Park SW, Brown MS, Goldstein JL. Combined analysis of oligonucleotide microarray data from transgenic and knockout mice identifies direct SREBP target genes. Proc Natl Acad Sci U S A. 2003 Oct 14;100(21):12027–32. PMID: 14512514.
18. Seo YK, Jeon TI, Chong HK, Biesinger J, Xie X, Osborne TF. Genome-wide localization of SREBP-2 in hepatic chromatin predicts a role in autophagy. Cell Metab. 2011 April 6;13(4):367–75. PMID: 21459322.
19. Mesmin B, Pipalia NH, Lund FW, Ramlall TF, Sokolov A, Eliezer D, Maxfield FR. STARD4 abundance regulates sterol transport and sensing. Mol Biol Cell. 2011 Nov;22(21):4004–15. PMID: 21900492.
20. Korytowski W, Rodriguez-Agudo D, Pilat A, Girotti AW. StarD4-mediated translocation of 7-hydroperoxycholesterol to isolated mitochondria: deleterious effects and implications for steroidogenesis under oxidative stress conditions. Biochem Biophys Res Commun. 2010 Jan 29;392(1):58–62. PMID: 20059974.
21. Yamada S, Yamaguchi T, Hosoda A, Iwawaki T, Kohno K. Regulation of human STARD4 gene expression under endoplasmic reticulum stress. Biochem Biophys Res Commun. 2006 May 19;343(4):1079–85. PMID: 16579971.
22. Rodriguez-Agudo D, Calderon-Dominguez M, Medina MA, Ren S, Gil G, Pandak WM. ER stress increases StarD5 expression by stabilizing its mRNA and leads to relocalization of its protein from the nucleus to the membranes. J Lipid Res. 2012 Dec;53(12):2708–15. PMID: 23053693.
23. Ye J, Rawson RB, Komuro R, Chen X, Dave UP, Prywes R, Brown MS, Goldstein JL. ER stress induces cleavage of membrane-bound ATF6 by the same proteases that process SREBPs. Mol Cell. 2000 Dec;6(6):1355–64. PMID: 11163209.
24. Zeng L, Lu M, Mori K, Luo S, Lee AS, Zhu Y, Shyy JY. ATF6 modulates SREBP2-mediated lipogenesis. EMBO J. 2004 Feb 25;23(4):950–8. PMID: 14765107.
25. Skiadas CC, Duan S, Correll M, Rubio R, Karaca N, Ginsburg ES, Quackenbush J, Racowsky C. Ovarian reserve status in young women is associated with altered gene expression in membrana granulosa cells. Mol Hum Reprod. 2012 July;18(7):362–71. PMID: 22355044.
26. Wei CL, Wu Q, Vega VB, Chiu KP, Ng P, Zhang T, Shahab A, Yong HC, Fu Y, Weng Z, Liu J, Zhao XD, Chew JL, Lee YL, Kuznetsov VA, Sung WK, Miller LD, Lim B, Liu ET, Yu Q, Ng HH, Ruan Y. A global map of p53 transcription-factor binding sites in the human genome. Cell. 2006 Jan 13;124(1):207–19. PMID: 16413492.
27. Bazuine M, Stenkula KG, Cam M, Arroyo M, Cushman SW. Guardian of corpulence: a hypothesis on p53 signaling in the fat cell. Clin Lipidol. 2009 April 1;4(2):231–43. PMID: 20126301.

28. ENCODE Project Consortium, Bernstein BE, Birney E, Dunham I, Green ED, Gunter C, Snyder M. An integrated encyclopedia of DNA elements in the human genome. Nature. 2012 Sept 6;489(7414):57–74. PMID: 22955616.
29. Chen YC, Meier RK, Zheng S, Khundmiri SJ, Tseng MT, Lederer ED, Epstein PN, Clark BJ. Steroidogenic acute regulatory-related lipid transfer domain protein 5 localization and regulation in renal tubules. Am J Physiol Renal Physiol. 2009 Aug;297(2):F380–8. PMID: 19474188.
30. Ishikawa T, Hwang K, Lazzarino D, Morris PL. Sertoli cell expression of steroidogenic acute regulatory protein-related lipid transfer 1 and 5 domain-containing proteins and sterol regulatory element binding protein-1 are interleukin-1beta regulated by activation of c-Jun N-terminal kinase and cyclooxygenase-2 and cytokine induction. Endocrinology. 2005 Dec;146(12):5100–11. PMID: 16123165.
31. Alpy F, Tomasetto C. Give lipids a START: the StAR-related lipid transfer (START) domain in mammals. J Cell Sci. 2005 July 1;118(Pt 13):2791–801. PMID: 15976441.
32. Alpy F, Tomasetto C. START ships lipids across interorganelle space. Biochimie. 2013 Sept 24. [Epub ahead of print] PMID: 24076129.
33. Elbadawy HM, Borthwick F, Wright C, Martin PE, Graham A. Cytosolic StAR-related lipid transfer domain 4 (STARD4) protein influences keratinocyte lipid phenotype and differentiation status. Br J Dermatol. 2011 March;164(3):628–32. PMID: 20969562.
34. Rodriguez-Agudo D, Ren S, Hylemon PB, Redford K, Natarajan R, Del Castillo A, Gil G, Pandak WM. Human StarD5, a cytosolic StAR-related lipid binding protein. J Lipid Res. 2005 Aug;46(8):1615–23. PMID: 15897605.
35. Rodriguez-Agudo D, Ren S, Hylemon PB, Montanez R, Redford K, Natarajan R, Medina MA, Gil G, Pandak WM. Localization of StarD5 cholesterol binding protein. J Lipid Res. 2006 June;47(6):1168–75. PMID: 16534142.
36. Strauss JF 3rd, Kishida T, Christenson LK, Fujimoto T, Hiroi H. START domain proteins and the intracellular trafficking of cholesterol in steroidogenic cells. Mol Cell Endocrinol. 2003 April 28;202(1–2):59–65. PMID: 12770731.
37. Romanowski MJ, Soccio RE, Breslow JL, Burley SK. Crystal structure of the Mus musculus cholesterol-regulated START protein 4 (StarD4) containing a StAR-related lipid transfer domain. Proc Natl Acad Sci U S A. 2002 May 14;99(10):6949–54. PMID: 12011453.
38. Thorsell AG, Lee WH, Persson C, Siponen MI, Nilsson M, Busam RD, Kotenyova T, Schuler H, Lehtio L. Comparative structural analysis of lipid binding START domains. PLoS One. 2011;6(6):e19521. PMID: 21738568.
39. Dikiy I, Ramlall TF, Eliezer D. ^{1}H, ^{13}C, and ^{15}N backbone resonance assignments of the L124D mutant of StAR-related lipid transfer domain protein 4 (StARD4). Biomol NMR Assign. 2013 Oct;7(2):245–8. PMID: 22918595.
40. Lorin A, Latourneau D, Lefebvre A, LeHoux JG, Lavigne P. ^{1}H, ^{13}C, and ^{15}N backbone chemical shift assignments of StAR-related lipid transfer domain protein 5 (STARD5). Biomol NMR Assign. 2013 April;7(1):21–4. PMID: 22392336.
41. Lavigne P, Najmanivich R, Lehoux JG. Mammalian StAR-related lipid transfer (START) domains with specificity for cholesterol: structural conservation and mechanism of reversible binding. Subcell Biochem. 2010;51:425–37. PMID: 20213553.
42. Chen BY, Honig B. VASP: a volumetric analysis of surface properties yields insights into protein-ligand binding specificity. PLoS Comput Biol. 2010 Aug 12;6(8):e1000881. PMID: 20814581.
43. Bose HS, Whittal RM, Ran Y, Bose M, Baker BY, Miller WL. StAR-like activity and molten globule behavior of StARD6, a male germ-line protein. Biochemistry. 2008 Feb 26;47(8):2277–88. PMID: 18211099.
44. Rodriguez-Agudo D, Ren S, Wong E, Marques D, Redford K, Gil G, Hylemon P, Pandak WM. Intracellular cholesterol transporter StarD4 binds free cholesterol and increases cholesteryl ester formation. J Lipid Res. 2008 July;49(7):1409–19. PMID: 18403318.

45. Latourneau D, Lorin A, Lefebvre A, Frappier V, Gaudreault F, Najmanovich R, Lavigne P, LeHoux JG. StAR-related lipid transfer domain protein 5 binds primary bile acids. J Lipid Res. 2012 Dec;53(12):2677–89. PMID: 23018617.
46. Latourneau D, Lorin A, Lefebvre A, Cabana J, Lavigne P, Lehoux JG. Thermodynamic and solution state NMR characterization of the binding of secondary and conjugated bile acids to STARD5. Biochim Biophys Acta. 2013 July 16;1831(11):1589–99. PMID: 23872533.
47. Latourneau D, Lefebvre A, Lavigne P, LeHoux JG. STARD5 specific ligand binding: comparison with STARD1 and STARD4 subfamilies. Mol Cell Endocrinol. 2013 May 22;371(1–2):20–5. PMID: 23337244.
48. Soccio RE, Breslow JL. Intracellular cholesterol transport. Arterioscler Thromb Vasc Biol. 2004 July;24(7):1150–60. PMID: 15130918.
49. Bose HS, Lingappa VR, Miller WL. Rapid regulation of steroidogenesis by mitochondrial protein import. Nature. 2002 May 2;417(6884):87–91. PMID: 11986670.
50. Clark BJ. The mammalian START domain protein family in lipid transport in health and disease. J Endocrinol. 2012 March;212(3):257–75. PMID: 21965545.
51. Borthwick F, Allen AM, Taylor JM, Graham A. Overexpression of STARD3 in human monocyte/macrophages induces an anti-atherogenic lipid phenotype. Clin Sci (Lond). 2010 June 22;119(7):265–72. PMID: 20491656.
52. Ning Y, Bai Q, Lu H, Li X, Pandak WM, Zhao F, Chen S, Ren S, Yin L. Overexpression of mitochondrial cholesterol delivery protein, StAR, decreases intracellular lipids and inflammatory factors secretion in macrophages. Atherosclerosis. 2009 May;204(1):114–20. PMID: 18945429.
53. Ning YX, Ren SL, Zhao FD, Yin LH. Overexpression of the steroidogenic acute regulatory protein increases the expression of ATP-binding cassette transporters in microvascular endothelial cells (bEnd.3). J Zhejiang Univ Sci B. 2010 May;11(5):350–6. PMID: 20443213.
54. Ning Y, Xu L, Ren S, Pandak WM, Chen S, Yin L. StAR overexpression decreases serum and tissue lipids in apolipoprotein E-deficient mice. Lipids. 2009 June;44(6):511–9. PMID: 19373502.
55. Skarnes WC, Rosen B, West AP, Koutsourakis M, Bushell W, Iyer V, Mujica AO, Thomas M, Harrow J, Cox T, Jackson D, Severin J, Biggs P, Fu J, Nefedov M, de Jong PJ, Stewart AF, Bradley A. A conditional knockout resource for the genome-wide study of mouse gene function. Nature. 2011 June 5;474(7351):337–42. PMID: 21677750.
56. Gomes C, Oh SD, Kim JW, Chun SY, Lee K, Kwon HB, Soh J. Expression of the putative sterol binding protein Stard6 gene is male germ cell specific. Biol Reprod. 2005 March;72(3):651–8. PMID: 15564601.
57. Ohta H, Tohda A, Nishimune Y. Proliferation and differentiation of spermatogonial stem cells in the w/wv mutant mouse testis. Biol Reprod. 2003 Dec;69(6):1815–21. PMID: 12890724.
58. Chang IY, Shin SY, Kim JW, Yu JM, Kim JS, Song PI, Yoon SP. The changed immunolocalization of START-domain-containing 6 (StarD6) during the development of testes in rat perinatal hypothyroidism. Acta Histochem. 2007;109(4):315–21. PMID: 17462719.
59. Golas A, Malek P, Piasecka M, Styrna J. Sperm mitochondria diaphorase activity–a gene mapping study of recombinant inbred strains of mice. Int J Dev Biol. 2010;54(4):667–73. PMID: 20209439.
60. Chang IY, Kim JH, Hwang G, Song PI, Song RJ, Kim JW, Yoon SP. Immunohistochemical detection of StarD6 in the rat nervous system. Neuroreport. 2007 Oct 8;18(15):1615–9. PMID: 17885612.
61. Chang IY, Kim JK, Lee SM, Kim JN, Soh J, Kim JW, Yoon SP. The changed immunoreactivity of StarD6 after pilocarpine-induced epilepsy. Neuroreport. 2009 July 1;20(10):963–7. PMID: 19434006.
62. Chang IY, Kim JH, Cho KW, Yoon SP. Acute responses of DNA repair proteins and StarD6 in rat hippocampus after domoic acid-induced excitotoxicity. Acta Histochem. 2013 April;115(3):234–9. PMID: 22883302.

63. Chang IY, Jeon YJ, Jung SM, Jang YH, Ahn JB, Park KS, Yoon SP. Does the StarD6 mark the same as the StAR in the nervous system? J Chem Neuroanat. 2010 Nov;40(3):239–42. PMID: 20609383.
64. Chang IY, Ohn T, Ko GS, Yoon Y, Kim JW, Yoon SP. Immunolocalization of steroidogenic acute regulatory protein-related lipid transfer (START) domain-containing proteins in the developing cerebellum of normal and hypothyroid rats. J Chem Neuroanat. 2012 Jan;43(1):28–33. PMID: 22024186.
65. Cappell KM, Sinnott R, Taus P, Maxfield K, Scarbrough M, Whitehurst AW. Multiple cancer testis antigens function to support tumor cell mitotic fidelity. Mol Cell Biol. 2012 Oct;32(20):4131–40. PMID: 22869527.
66. Ashrafi K, Chang FY, Watts JL, Fraser AG, Kamath RS, Ahringer J, Ruvkun G. Genome-wide RNAi analysis of *Caenorhabditis elegans* fat regulatory genes. Nature. 2003 Jan 16;421(6920):268–72. PMID: 12529643.
67. Li S, Armstrong CM, Bertin N, Ge H, Milstein S, Boxem M, Vidalain PO, Han JD, Chesneau A, Hao T, Goldberg DS, Li N, Martinez M, Rual JF, Lamesch P, Xu L, Tewari M, Wong SL, Zhang LV, Berriz GF, Jacotot L, Vaglio P, Reboul J, Hirozane-Kishikawa T, Li Q, Gabel HW, Elewa A, Baumgartner B, Rose DJ, Yu H, Bosak S, Sequerra R, Fraser A, Mango SE, Saxton WM, Strome S, Van Den Heuvel S, Piano F, Vandenhaute J, Sardet C, Gerstein M, Doucette-Stamm L, Gunsalus KC, Harper JW, Cusick ME, Roth FP, Hill DE, Vidal M. A map of the interactome network of the metazoan *C. elegans*. Science. 2004 Jan 23;303(5657):540–3. PMID: 14704431.
68. Urano F, Calfon M, Yoneda T, Yun C, Kiraly M, Clark SG, Ron D. A survival pathway for *Caenorhabditis elegans* with a blocked unfolded protein response. J Cell Biol. 2002 Aug 19;158(4):639–46. PMID: 12186849.

Chapter 8
Steroidogenic Acute Regulatory Protein-related Lipid Transfer (START) Proteins in Non-vesicular Cholesterol Transport

David B. Iaea, Shu Mao and Frederick R. Maxfield

Abstract Lipid transfer proteins play an important role in non-vesicular transport of sterols, phospholipids, and sphingolipids among intracellular membranes to maintain the proper sterol and lipid distribution. The steroidogenic acute regulatory protein-related lipid transfer (START) domain family are defined by a conserved 210 amino acid sequence that folds into an α/β helix-grip structure containing a hydrophobic pocket for ligand binding. The mammalian START proteins bind a variety of ligands, including cholesterol, phospholipids, sphingolipids, and bile acids with putative roles in non-vesicular lipid transport, tumor suppression, and thioesterase activity. However, the functions of many START proteins have yet to be well characterized. Recent studies have focused on determining the cell type distribution and expression profile of the START proteins, examining the ligand specificity and directionality of transport and characterizing disease states that may be associated with deregulation of START proteins. This chapter will summarize current findings regarding the physiologic and pathological roles of the START proteins in non-vesicular lipid transport.

Abbreviations

ACAT	acyl-CoA:cholesterol acyl-transferase
CHO	Chinese hamster ovary
DHE	dehydroergosterol
ER	endoplasmic reticulum
ERC	endocytic recycling compartment
EST	expressed sequence tag
HER2	human epidermal growth factor receptor 2
HMGR	HMG-CoA reductase
INSIG 1–2	insulin-induced genes 1 and 2
LDL	low density lipoproteins
LE/LY	late endosome and lysosomes
MENTAL	MLN64-NH2-terminal

F. R. Maxfield, Ph.D. (✉) · D. B. Iaea · S. Mao, Ph.D.
Department of Biochemistry, Weill Cornell Medical College,
1300 York Ave, New York, NY, 10065, USA
e-mail: frmaxfie@med.cornell.edu

B. J. Clark, D. M. Stocco (eds.), *Cholesterol Transporters of the START Domain Protein Family in Health and Disease,* DOI 10.1007/978-1-4939-1112-7_8,

MLN64	metastatic axillary lymph node 64 kDa protein
NPC	Niemann-Pick C
ORP	OSBP-related protein
OSBP	oxysterol binding protein
PC	phosphatidylcholine
PCTP	phosphatidylcholine transfer protein
PE	phosphatidylethanolamine
PH	pleckstrin homology
PM	plasma membrane
PS	phosphatidylserine
S1P and S2P	site 1 and site 2 proteases
SCAP	SREBP-cleavage-activating protein
SREBP-2	sterol regulatory element-binding protein-2
STARD	START-related domain
START	steroidogenic acute regulatory protein-related lipid transfer
VAP	vesicle-associated membrane protein-associated protein

Introduction

Our laboratory's initial interest was the basic cellular process of receptor-mediated endocytosis [1], and our main research tool has been quantitative fluorescence microscopy. We initially used low-density lipoproteins (LDL) as a fluorescent probe to study endocytic pathways because they could be labeled brightly, which was very useful given the cameras available in the 1980s and 1990s. Later, in collaboration with Ira Tabas at Columbia University, we began to explore the role of lipoproteins in the formation of macrophage foam cells [2]. This led to investigations into how LDL-derived cholesterol gets from late endosomes and lysosomes (LE/LY) into the endoplasmic reticulum (ER; [3]), followed by a realization that this process was not well understood. We then explored the use of a fluorescent sterol, dehydroergosterol (DHE), that had been studied extensively by Schroeder's group [4, 5], and we began to study its intracellular transport [6]. It became obvious that there was extensive non-vesicular intracellular transport of sterols in mammalian cells, and we wanted to understand what the carriers might be. Jan Breslow, at Rockefeller University, discussed STARD4 with us because of his interest in this as an SREBP2-regulated protein, and we began a collaboration to investigate the possible role of STARD4 and STARD5 in sterol transport. Our initial studies were not successful, mainly because the microscopy tools available at the time were inadequate. With better instruments and a better understanding of the cell biology, we were able to publish two papers to date on the cellular role of STARD4 in sterol transport and regulation [7, 8]. We continue to explore this fascinating but challenging family of proteins.

Sterols are ubiquitous components of cell membranes in eukaryotes. Among the major lipid components of eukaryotic membranes, sterols have a unique chemical structure. Cholesterol, a 27 carbon molecule, contains a single hydroxyl as the only polar component, four planar rings and a short alkyl chain [9]. Its structure is in contrast with most phospholipids, which are composed of large polar headgroups and long hydrocarbon chains. Cholesterol is heterogeneously distributed among cellular organelles with ~60 % of total cellular cholesterol in the plasma membrane (PM) of mammalian cells, while relatively low amounts are maintained in the mitochondria and ER, the latter being the site of cholesterol biosynthesis and storage [10–14]. In several cell types, the endocytic recycling compartment (ERC) has been shown to be a major pool of intracellular cholesterol [6, 15].

Cellular cholesterol content has major effects on the biophysical properties of membranes [16], which can alter the functional properties of membrane proteins [17, 18]. Therefore, maintaining the proper distribution of cholesterol among cellular membranes is required for homeostasis. As discussed in this chapter, sterol transport among organelles plays a key role in the cell's mechanisms for maintaining cholesterol homeostasis. At the cellular level, excess cholesterol can be regulated rapidly by esterification by acyl-CoA:cholesterol acyl-transferase (ACAT) followed by storage in lipid droplets. The rate-limiting step in esterification of cholesterol by ACAT is delivery of cholesterol to the ER, which can be assisted by various lipid transfer proteins [15, 19]. At the level of the whole organism, cholesterol is metabolized in the liver to bile acids, and excretion of bile is the major pathway for elimination of cholesterol. Dysregulation of cholesterol homeostasis is an important contributor to diseases such as atherosclerosis and some lysosomal storage disorders, and it can play a role in some cancers as well [20–22].

Cellular cholesterol levels are tightly regulated by coordinated biosynthetic and efflux pathways. Mammalian cells obtain cholesterol by two primary mechanisms: uptake from lipoproteins and *de novo* biosynthesis. Figure 8.1 illustrates pathways of intracellular sterol transport. Alterations in the cholesterol content of the ER causes profound changes in the expression of genes involved in the biosynthesis and uptake of cholesterol through the actions of the Insig-SCAP-SREBP2 proteins [23]. Endogenous cellular cholesterol biosynthesis occurs in the ER, which contains the major rate-limiting enzyme in cholesterol biosynthesis, HMG-CoA reductase (HMGR). HMGR expression is transcriptionally regulated by the sterol regulatory element-binding protein-2 (SREBP2) transcription factor [24]. When cellular sterols are abundant, SREBPs are retained in the ER membrane in a complex with the cholesterol-sensing protein, SREBP-cleavage-activating protein (SCAP), and the ER retention proteins insulin-induced genes 1 and 2 (INSIG 1–2) [11].

When cholesterol content in the ER is depleted, SCAP undergoes a conformational change and is released from Insig, allowing the SREBP2-SCAP complex to translocate to the Golgi apparatus [13]. In the Golgi, SREBP2 is proteolytically processed by the site 1 and site 2 proteases (S1P and S2P) to release the 50 kDa amino (N)-terminal transcription factor that translocates to the nucleus to activate the expression of genes involved in sterol biosynthesis, uptake, and metabolism [25]. Therefore, changes in cholesterol levels in membranes such as the PM must

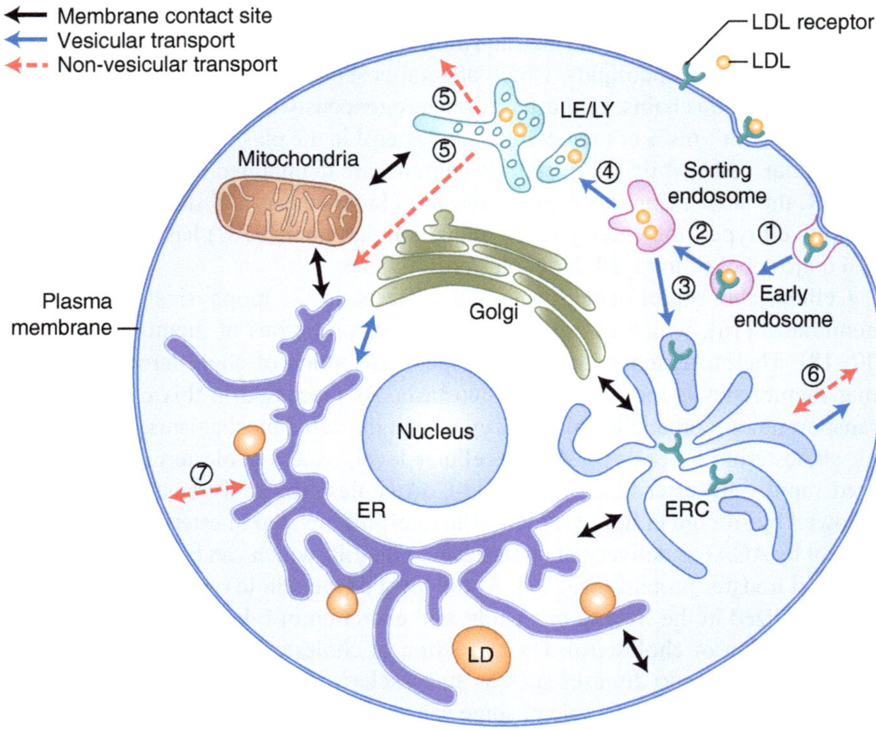

Fig. 8.1 1, 2 LDL particles containing cholesterol-esters and free cholesterol are internalized by clathrin-mediated endocytosis of the LDL receptor [1]. 3 The low pH in early endosomes causes the LDL to be released from its receptor, and the empty LDL receptor returns to the cell surface for further rounds of endocytosis. 4 The LDL is retained in the early sorting endosomes, which mature into late endosomes where it encounters acid hydrolases including lysosomal acid lipase, which hydrolyzes the cholesteryl esters in the core of the LDL particles. 5 The free cholesterol is transported out of LE/LY and delivered to other cellular membranes including the plasma membrane and ER by primarily non-vesicular mechanisms. 6 Cholesterol in the plasma membrane can traffic to the ERC and back by both vesicular and non-vesicular mechanisms. 7 Cholesterol transport from the plasma membrane to the ER informs the homeostatic machinery about the free cholesterol levels in the cell. *LDL* low density lipoproteins, *LE/LY* late endosome and lysosomes, *ER* endoplasmic reticulum, *ERC* endocytic recycling compartment

be sensed in the ER to maintain cholesterol homeostasis. Studies of lipid trafficking from the PM generally find extremely low levels of delivery to the ER [26], indicating that non-vesicular transport must be required for effective movement of cholesterol from the PM [19] to the ER [1].

This poses an interesting question in that, in order for cholesterol levels sensed in the ER to reflect the cholesterol distribution and content in other organelles, such as the PM and endosomes, there must be a mechanism for rapid redistribution of cholesterol. Several lines of evidence indicate that non-vesicular transport plays an important role in maintaining the correct distribution of cholesterol among organelles. Lipid transfer proteins play an important role in non-vesicular transport of sterols and lipids among intracellular organelles [27]. They also play an important role in sensing abundance of lipids and regulating cellular physiology [28]. Newly synthesized cholesterol in the ER is rapidly trafficked to the PM, by non-vesicular mechanisms [29–31]. A fluorescent sterol, DHE, showed a distribution in Chinese hamster ovary (CHO) fibroblasts similar to cholesterol [6]. When the DHE in the ERC was photobleached, much of it was replaced by DHE from elsewhere in the cell (mainly the PM) within a few minutes [6]. Most of this transport of DHE was adenosine triphosphate (ATP)-independent, so it must have been by a non-vesicular process. Additionally, following receptor mediated endocytosis of LDL, free cholesterol is generated by hydrolysis of cholesterol esters in the LE/LY compartments [32]. The cholesterol that is released from the LE/LY is distributed throughout the cell, including the PM [19] and the ER [1, 33]. Since there are not major vesicular trafficking routes from the LE/LY to either the PM or the ER, it is likely that much of this transport is non-vesicular.

There are several protein families that are classified as sterol transfer proteins, which have been shown to be capable of transferring sterols between membranes [34]. Sterol transfer proteins can interact directly with membrane compartments to extract sterol and then diffuse in complex with the sterol to acceptor membranes [35]. In some cases, such as ER-PM contact sites, the membranes of two organelles are held in close proximity by protein complexes, and this can facilitate rapid exchange of sterol between the membranes by reducing the distance that the sterol:protein complex must travel. This mechanism would also provide a means for targeted delivery of sterol from donor to acceptor membranes. Two major gene families for lipid transfer proteins that have been implicated in such trafficking are the steroidogenic acute regulatory protein (StAR)-related lipid-transfer (START) domain family [36] and the oxysterol-binding protein (OSBP) and OSBP-related protein (ORP) family, which have been discussed in detail elsewhere [37, 38]. This chapter will summarize the current knowledge of the physiologic roles of the START proteins in non-vesicular lipid transport.

Sterol Transport Proteins and the START Domain Family

The START domain is defined by a 210 amino acid sequence that folds into an α/β helix-grip structure containing a hydrophobic pocket for ligand binding [39, 40]. Database analyses have identified START domains in the genomes of plants, bacteria, protists, and animals, but they have not been identified in either archaea or yeast. Interestingly, START domains are most abundant in plants and are found in proteins that contain a homeodomain, suggesting a role in transcriptional regulation [41]. This unique protein architecture of the START-homeodomain is found only in plants. However, the presence of a START domain in conjunction with other motifs, to form multi-domain proteins, is common in various phyla and permits additional protein functionality such as enzymatic activity, signaling, and protein localization [42].

The mammalian START domain protein family is composed of 15 members that group into six subfamilies based on domain architecture and ligand specificity [42]. In general terms, there are the cholesterol/oxysterol binding proteins (STARD1/3 subfamily), soluble proteins (STARD4/5/6 subfamily), phospholipid- and sphingolipid-binding proteins (STARD2 [phosphatidylcholine transfer protein, PCTP]/7/8/10/11 subfamily), putative Rho-GTPase signaling (STARD8/12/13 subfamily), thioesterase activity containing proteins (STAR4/15 subfamily), and the STARD9 subfamily composed of a single member.

Mechanism of START-mediated Sterol Transport

The crystal structures of human STARD3/metastatic axillary lymph node 64 kDa protein (MLN64) and murine STARD4 were among the first of the START proteins to be described [39, 43]. The structures of these proteins showed an α/β helix-grip structure composed of nine anti-parallel β sheets flanked at the amino- and carboxyl-termini by α helices. The core of this protein forms a hydrophobic pocket that would accommodate a single sterol molecule. The suggested mechanisms for START domain cholesterol transfer has been described elsewhere [44, 45]. In brief, structural and biophysical studies have proposed two models: Movement of the carboxyl termini helix leads to a molten globule transition to facilitate cholesterol absorption and release [44, 46]. Alternatively, molecular dynamics studies have proposed that movement of the omega-1 loop, following membrane binding, may be sufficient for activity [45]. Comparing the structure of several sterol transfer proteins, demonstrates that they each contain a hydrophobic ligand pocket capable of binding a single sterol molecule [43, 47, 48]. In several of these proteins, the sterol-binding pocket is under "gated" regulation that could open and close upon interaction with the lipid bilayer, as suggested by molecular dynamic studies, to facilitate sterol/lipid absorption and delivery to specific membranes. To date, several crystal structures of START family members have been reported, confirming that the helix-grip structure is maintained across five mammalian START subfamilies [49, 50].

The Soluble Sterol Transporters STARD4 Subfamily

The STARD4 subfamily is composed of STARD4, STARD5, and STARD6. This subfamily is most closely related to the STARD1/3 subfamily with approximately 25 % sequence identity in the START domains [51]. STARD4 was identified as a novel expressed sequence tag (EST) in a microarray designed to identify genes that are transcriptionally regulated by cholesterol in the liver. Briefly, mice that were fed a high cholesterol diet had reduced STARD4 messenger ribonucleic acid (mRNA) levels in comparison to control animals. Expression of STARD4 in 3T3-L1 fibroblasts and a human monocytic cell line derived from an acute monocytic leukemia patient (THP-1) was suppressed by sterol overloading and enhanced by statin treatment [52]. These findings are consistent with STARD4 being transcriptionally regulated by cellular cholesterol levels via SREBP-2 [53]. Subsequently, STARD5 and STARD6 were identified by BLAST search of the whole genome and EST databases [51]. Analysis of the protein-coding sequence for the STARD4 family indicates a protein of ~25 kDa that is entirely composed of the START domain without a membrane targeting sequence. The cytosolic distribution of these START proteins has led to increased interest and speculation regarding the role of these proteins in cholesterol trafficking and homeostasis. To understand the role of this START protein subfamily, we will discuss the properties and proposed functions of its members.

STARD4: Cholesterol Trafficking Among Intracellular Organelles

STARD4 is widely expressed in a variety of tissues [51] and was detected by immunohistochemistry in human hepatocytes and Kupffer cells [52]. The subcellular localization of STARD4 is difficult to ascertain. As a soluble cytoplasmic protein with no obvious targeting motif, it may be distributed throughout the cytoplasm. Immunolocalization requires permeabilization of cells with detergents, which can release many cytoplasmic proteins even in fixed cells and can also disrupt the membranes to which a protein such as STARD4 may bind. It has been reported that STARD4 can be seen associated with the ER by immunolocalization [52], but the fraction that is associated with the ER in the intact cell is difficult to determine.

Several lines of evidence, both in cells and in vitro, show that STARD4 can transfer cholesterol efficiently. STARD4 can transfer sterol between liposomes in a process that depends on the composition of the donor and acceptor membranes [7]. Analysis of the STARD4 surface revealed a region enriched in basic residues near the sterol-binding pocket that facilitates STARD4's interaction with negatively charged membranes [7]. Mutations in the basic patch markedly decreased the sterol transfer rate of STARD4, indicating that the interaction with anionic lipids is required for maximal activity. This finding is of particular interest since the cytoplasmic leaflets of the PM, the ERC, and ER are highly enriched in anionic lipids, particularly phosphatidylserine [54]. Additionally, the sterol transfer rate of STARD4 was increased by threefold when acceptor liposomes were enriched in unsaturated lipids [7]. Since the ER is more highly enriched in unsaturated lipids than other organelles [55], STARD4-mediated delivery of sterol to the ER could be enhanced by the abundance of unsaturated lipids.

Several experimental results support the hypothesis that STARD4 plays an important role in delivery of cholesterol to the ER. In studies with isolated microsomes, cholesterol esterification by ACAT1, an ER enzyme, is stimulated by purified STARD4 [52]. Overexpression of STARD4 in mouse hepatocytes leads to an increase in cholesterol esterification by ACAT [52]. The role of STARD4 in transporting cholesterol to the ER is further supported by a recent study of STARD4 in U2OS human osteosarcoma cells [7]. STARD4 overexpression increased cholesterol ester levels. STARD4 silencing by small interfering RNA (siRNA) attenuated cholesterol-mediated regulation of SREBP-2 activation, while STARD4 overexpression amplified sterol sensing by the SCAP/SREBP-2 proteins in the ER. Silencing STARD4 increased the cellular cholesterol levels, presumably as a result of the slower delivery of cholesterol to the ER. Interestingly, this effect of reduced STARD4 expression on cholesterol levels could be reversed by microinjection of methyl-β-cyclodextrin, a cyclic sugar that nonselectively exchanges cholesterol among membranes. These data suggest that delivery of cholesterol to the sterol-sensing membranes of the ER is an important function of STARD4.

STARD4 also plays a role in sterol transport to other organelles. When U2OS cells are incubated with the fluorescent sterol, DHE, the ERC is the major intracellular organelle that is labeled [6]. Fluorescence recovery after photobleaching the ERC allows measurement of the rate of transport of DHE back into the ERC, and overexpression of STARD4 significantly increased the transport of DHE into the ERC [7]. Knockdown of STARD4 in HepG2 cells resulted in decreased sterol transport to the ERC as well as reduced ER-associated cholesterol and cholesterol esters [8]. Again, microinjection of methyl-β-cyclodextrin into the cells had a similar effect to overexpression of STARD4. Transient expression of STARD4 in Cos-1 cells resulted in an increase in steroidogenesis though at a lower efficiency than StAR and STARD3 [53, 56], suggesting that STARD4 may be able to deliver cholesterol to mitochondria. This indicates that there may be some overlap in the function of the StAR sterol transporters.

Increased cholesterol transport to the ER may result in ER stress. While it is well established that STARD4 is transcriptionally regulated by SREBP-2, STARD4 mRNA was also found to be increased following transient treatment with tunicamycin, a small molecule activator of ER stress [53, 57]. While the functional relevance of increased STARD4 mRNA following ER stress is unknown, transient increases in STARD4 under disease conditions associated with ER stress may affect cholesterol homeostasis by increasing cholesterol delivery to the ER, which would increase both SREBP-2 processing and cholesterol esterification.

Recently, a homozygous STARD4 knockout mouse has been described [58]. Interestingly, STARD4 knockout mice develop normally and do not present with a distinct lipid phenotype; plasma and hepatic lipid profiles of STARD4 knockout mice were similar to wild-type mice. Female STARD4 knockout mice did have decreased phospholipid and cholesterol content in their gallbladder [58]. Statin treatment did not alter the hepatic or plasma lipid profiles of wild-type or STARD4 knockout mice. The mRNA levels of various START proteins were not significantly altered in STARD4 knockout mice, indicating that expression of other START proteins is not responsive to loss of STARD4 [58]. This likely indicates that there is redundancy for sterol transport, beyond the START family, in mammalian cells. It remains unclear whether there are stresses—dietary or disease specific—that may be exacerbated by loss of STARD4.

STARD5 and ER Stress

STARD5 has been reported to bind cholesterol as well as 25-hydroxycholesterol, and it is widely expressed in a variety of tissues with the highest expression in liver and kidney [51, 59]. In the liver, STARD5 is localized solely to the Kupffer cells [60]. STARD5 does not transfer cholesterol to mitochondria in cells or in vitro [61]. Expression of STARD5 in hepatocytes results in a threefold increase in free cholesterol levels, particularly in the Golgi region and possibly the ER [60]. As a sterol transporter, it has been suggested that STARD5 may shuttle cholesterol among organelles [60, 62]. However, recent structural studies of STARD5 have shown that it does not bind cholesterol but rather cholic acid, a precursor in bile acid biogenesis [63]. With this new finding, further studies will be required to identify the biologically relevant substrates of STARD5. It is reported that unlike STARD4, STARD5 overexpression does not result in an increase in cholesterol ester formation, but increases free cholesterol levels, indicating distinct roles and functions for these sterol transporters in cholesterol homeostasis [7, 59, 61]. Additionally, overexpression of STARD5 in macrophages increased SREBP2 mRNA levels, which may result in increase in cholesterol biosynthesis [64]. It is likely that STARD5 participates in some aspect of cholesterol regulation, but its precise role is uncertain.

STARD5 mRNA levels are increased following treatment with agents that promote ER stress, such as thapsigargin in 3T3 cells or cholesterol overloading in J774 macrophages [53, 62]. The precise function of STARD5 under ER stress is not known, but chronic ER stress is associated with several metabolic disease states including type 2 diabetes and cancer [62, 65]. In a diabetic mouse model, STARD5 mRNA and protein steady state levels were significantly increased, and free cholesterol levels were elevated compared to control animals [62].

STARD6: Cholesterol Transport to the Mitochondria

STARD6 was identified as being predominantly expressed in testes and was later shown to be solely localized to the germ lines, with highest expression in spermatids [51]. However, the function of STARD6 in spermatids remains unknown. Recent work has suggested a role for STARD6 in sperm motility and quality as it is a gene required for mitochondrial nicotinamide adenine dinucleotide hydrogen (NADH)-dependent dehydrogenase activity [66]. This finding is in agreement with previous reports that STARD6 transports cholesterol to the mitochondria as effectively as StAR and STARD3 [46, 61]. These findings suggest that STARD6 may function to deliver cholesterol to the mitochondria in male germ cells, but the exact functions remains to be determined. Like the other members of the STARD4 subfamily, STARD6 lacks an organelle-targeting sequence, so further work is required to investigate the role of STARD6 in cholesterol transport to specific organelles (mitochondria, ER, or PM) in male germ cells [42].

The Oxysterol-Cholesterol-Binding Proteins STARD1/3

StAR and STARD3 are similar in that both START domains are selective for cholesterol and are targeted to specific membrane compartments [67]. StAR is the prototypic and founding member of the START domain protein family. StAR is primarily expressed in the adrenal and gonads. Among the START domain family, StAR contains a classical amino-terminal-targeting sequences that directs it to the mitochondria [68]. Recent biochemical studies indicate that StAR may form a functional complex with a cholesterol transfer channel where it facilitates cholesterol transport from the outer to inner mitochondrial membrane [69, 70]. However, the precise molecular mechanism for StAR cholesterol transfer across the membranes of mitochondria is not known although various studies have provided insights into this process. Greater details regarding the current model of StAR and its mechanism of action can be found in Chaps. 2 and 3 as well as recent reviews [36, 44, 68].

STARD3 and Cholesterol Efflux from LE/LY

STARD3/MLN64 is a multi-pass transmembrane protein with an MLN64-NH2-terminal (MENTAL) domain that localizes to late endosomes with the START domain orientated into the cytosol at the carboxy terminus [71, 72]. The localization of MLN64 to late endosomes has suggested a role in cholesterol efflux from this compartment as well as a potential role in Niemann-Pick C (NPC) disease, which has been discussed in preceding chapters [27, 73]. NPC disease is a lipid storage disorder caused by mutation in either NPC1 or NPC2 that results in accumulation of free cholesterol in LE/LY, leading to systemic and neurological disorders [20]. Briefly, following endocytosis of LDL and generation of free cholesterol by lysosomal acid lipase in LE/LY, NPC2 binds cholesterol and transfers it to the cholesterol-binding domain of NPC1, a transmembrane protein in the limiting membrane of the lysosome. NPC1 transfers cholesterol out of the lysosome by an unknown mechanism [74, 75]. STARD3 has been proposed to participate in the efflux of cholesterol from the LE/LY, but the mechanism for this is unclear [73]. It has been proposed that STARD3 plays a role in cholesterol transport from the LE/LY to mitochondria for steroidogenesis [75]. However, the exact function of STARD3/MLN64 in cholesterol trafficking remains unclear [76].

The Phospho/Sphingolipid-Binding START Proteins

This subfamily of the START protein is composed of the phosphatidylcholine (PC)-binding STARD2 (PCTP), STARD7, and STARD10 as well as STARD11, which binds ceramide [42]. Of the STARD2 family, STARD2/PCTP and STARD11/CERT have been extensively studied and reviewed [50, 77, 78]. Purified PCTP was shown to selectively transport PC between membranes in cells and in vitro [79, 80]. A potential function for STARD2 is to facilitate PC delivery to the hepatic canalicular and alveolar membrane for incorporation into bile acids and lung surfactant synthesis, respectively [81]. The crystal structure of PCTP bound to PC demonstrates the characteristic helix-grip fold and hydrophobic tunnel of the START family. Interestingly, the PC head group provides ligand specificity, and the hydrophobic tunnel can accommodate PC regardless of acyl chain saturation [49].

STARD7 was originally identified as a transcript overexpressed in placental choriocarcinoma cells [82]. It shares ~25 % sequence identity with PCTP and selectively transfers PC but not phosphatidylserine (PS), phosphatidylethanolamine (PE), or sphingomyelin between membranes in vitro [83]. Similarly, STARD10 was identified as a phosphoprotein overexpressed in tumors of the human epidermal growth factor receptor 2 (HER2) transgenic mouse [84]. Purified STARD10 transfers PC and PE in vitro and in vivo but displays some ligand specificity toward PC in comparison to PE [85].

STARD7 and PC Distribution

A variant of STARD7, STARD7–1, was identified by protein database analysis and was found to have an extended amino terminus that forms an amphipathic helix to localize the protein to the mitochondria [83]. In a mouse hepatoma cell line, overexpression of STARD7 resulted in an increase in mitochondrial PC levels while silencing using siRNA did not alter PC levels in the mitochondria. STARD7 may function at the cytoplasmic leaflet of the outer mitochondrial membrane to regulate mitochondrial PC levels [83]. STARD7 is highly expressed in proliferating cells including lung, colon, and liver cells and may function to deliver PC for mitochondrial biogenesis, function, and homeostasis.

STARD11/CERT and Ceramide Transport

STARD11, more commonly known as ceramide transfer protein (CERT), is a ceramide transport protein that acts to transport ceramide from the ER to the Golgi [86], where it is converted to sphingomyelin and glucosylceramide [87]. Unlike the other members of the STARD2 family, STARD11/CERT has additional structural domains that target the START domain to selective membrane compartments for cellular function. CERT has an amino terminal pleckstrin homology (PH) domain, a central Two phenylalanines in an acidic tract (FFAT) motif, and a carboxy terminal START domain. The PH domain and the FFAT motif act to target CERT to either the Golgi by binding of the PH domain to the phosphoinositide, PI4P, or to the ER by interaction with the ER resident protein vesicle-associated membrane protein-associated protein (VAP), via the FFAT motif [72]. The phosphorylation of CERT is proposed to enhance membrane interaction by allowing the PH and FFAT motifs to interact with their respective membrane targets in order to position the START domain for ceramide extraction and delivery [78, 86]. The structure of the START domain of CERT has been determined and confirms the helix-grip fold and hydrophobic pocket that would facilitate ceramide binding [50].

Summary

As illustrated in Fig. 8.1, non-vesicular lipid trafficking plays an essential role in the maintenance of cellular lipid distribution. Recent advances in biochemistry and cell biology have led to the description of several basic properties, including the rate of sterol transport. Unfortunately, very little is understood about the molecular mechanisms involved in non-vesicular transport. Several proteins have been identified as sterol transporters including the START and ORP families. Of the mammalian START domain proteins, the STARD1/3 and STARD4 families are involved in sterol transport. While the biological function of StAR is well established, further work remains to elucidate the functional relevance of STARD3/4/5/6. Interestingly, STARD4 and STARD5 have been proposed to function at the ER although with distinct roles. Knockout animals of STARD4 were viable and did not have altered lipid profiles. However, it is possible that a phenotype following loss of STARD4 could potentially be exacerbated by dietary or disease related stresses. The STARD2/PCTP subfamily functions in a variety of pathways and transports several different lipids. It is important to continue to examine START protein expression and function in both normal and disease states in order to establish and distinguish their biological significance in lipid homeostasis.

Acknowledgements Work in the Maxfield laboratory is supported by National Institutes of Health (NIH) grant R37-DK27083.

References

1. Mukherjee S, Ghosh RN, Maxfield FR. Endocytosis. Physiol Rev. 1997;77(3):759–803.
2. Myers JN, Tabas I, Jones NL, Maxfield FR. Beta-very low density lipoprotein is sequestered in surface-connected tubules in mouse peritoneal macrophages. J Cell Biol. 1993;123(6 Pt 1):1389–402.
3. Skiba PJ, Zha X, Maxfield FR, Schissel SL, Tabas I. The distal pathway of lipoprotein-induced cholesterol esterification, but not sphingomyelinase-induced cholesterol esterification, is energy-dependent. J Biol Chem. 1996;271(23):13392–400.
4. McIntosh AL, Atshaves BP, Huang H, Gallegos AM, Kier AB, Schroeder F. Fluorescence techniques using dehydroergosterol to study cholesterol trafficking. Lipids. 2008;43(12):1185–208.
5. Schroeder F, Barenholz Y, Gratton E, Thompson TE. A fluorescence study of dehydroergosterol in phosphatidylcholine bilayer vesicles. Biochemistry. 1987;26(9):2441–8.
6. Hao M, Lin SX, Karylowski OJ, Wustner D, McGraw TE, Maxfield FR. Vesicular and non-vesicular sterol transport in living cells. The endocytic recycling compartment is a major sterol storage organelle. J Biol Chem. 2002;277(1):609–17.

7. Mesmin B, Pipalia NH, Lund FW, Ramlall TF, Sokolov A, Eliezer D, et al. STARD4 abundance regulates sterol transport and sensing. Mol Biol Cell. 2011;22(21):4004–15.
8. Garbarino J, Pan M, Chin HF, Lund FW, Maxfield FR, Breslow JL. STARD4 knockdown in HepG2 cells disrupts cholesterol trafficking associated with the plasma membrane, ER, and ERC. J Lipid Res. 2012;53(12):2716–25.
9. van Meer G. Cellular lipidomics. EMBO J. 2005;24(18):3159–65.
10. Ikonen E. Cellular cholesterol trafficking and compartmentalization. Nat Rev Mol Cell Biol. 2008;9(2):125–38.
11. Radhakrishnan A, Goldstein JL, McDonald JG, Brown MS. Switch-like control of SREBP-2 transport triggered by small changes in ER cholesterol: a delicate balance. Cell Metab. 2008;8(6):512–21.
12. Maxfield FR, Menon AK. Intracellular sterol transport and distribution. Curr Opin Cell Biol. 2006;18(4):379–85.
13. Blanchette-Mackie EJ, Pentchev P. Cholesterol distribution in Golgi, lysosomes and endoplasmic reticulum. In: Chang TY, Freeman D, editors. Intracellular cholesterol trafficking. US: Springer; 1998. S. 53–73.
14. Mobius W, van Donselaar E, Ohno-Iwashita Y, Shimada Y, Heijnen HFG, Slot JW, et al. Recycling compartments and the internal vesicles of multivesicular bodies harbor most of the cholesterol found in the endocytic pathway. Traffic. 2003;4(4):222–31.
15. Liscum L, Munn NJ. Intracellular cholesterol transport. Biochim Biophys Acta. 1999;1438(1):19–37.
16. Huang J, Feigenson GW. A microscopic interaction model of maximum solubility of cholesterol in lipid bilayers. Biophys J. 1999;76(4):2142–57.
17. Pani B, Ong HL, Liu X, Rauser K, Ambudkar IS, Singh BB. Lipid rafts determine clustering of STIM1 in endoplasmic reticulum-plasma membrane junctions and regulation of store-operated Ca2+ entry (SOCE). J Biol Chem. 2008;283(25):17333–40.
18. Galan C, Woodard GE, Dionisio N, Salido GM, Rosado JA. Lipid rafts modulate the activation but not the maintenance of store-operated Ca(2+) entry. Biochim Biophys Acta. 2010;1803(9):1083–93.
19. Das A, Goldstein JL, Anderson DD, Brown MS, Radhakrishnan A. Use of mutant 125I-Perfringolysin O to probe transport and organization of cholesterol in membranes of animal cells. Proc Natl Acad Sci U S A. 2013;110(26):10580–5.
20. Rosenbaum AI, Maxfield FR. Niemann-Pick type C disease: molecular mechanisms and potential therapeutic approaches. J Neurochem. 2011;116(5):789–95.
21. Siperstein MD, Fagan VM. Deletion of the cholesterol-negative feedback system in liver tumors. Cancer Res. 1964;24:1108–15.
22. Tabas I. Apoptosis and plaque destabilization in atherosclerosis: the role of macrophage apoptosis induced by cholesterol. Cell Death Differ. 2004;11(Suppl 1):2–6.
23. Goldstein JL, DeBose-Boyd RA, Brown MS. Protein sensors for membrane sterols. Cell. 2006;124(1):35–46.
24. Brown MS, Goldstein JL. Cholesterol feedback: from Schoenheimer's bottle to Scap's MELADL. J Lipid Res. 2009;50(Suppl):5–27.
25. Horton JD, Shah NA, Warrington JA, Anderson NN, Park SW, Brown MS, et al. Combined analysis of oligonucleotide microarray data from transgenic and knockout mice identifies direct SREBP target genes. Proc Natl Acad Sci U S A. 2003;100(21):12027–32.
26. Wustner D, Mondal M, Tabas I, Maxfield FR. Direct observation of rapid internalization and intracellular transport of sterol by macrophage foam cells. Traffic. 2005;6(5):396–412.
27. Mesmin B, Maxfield FR. Intracellular sterol dynamics. Biochim Biophys Acta. 2009;1791(7):636–45.
28. Li X, Rivas MP, Fang M, Marchena J, Mehrotra B, Chaudhary A, et al. Analysis of oxysterol binding protein homologue Kes1p function in regulation of Sec14p-dependent protein transport from the yeast golgi complex. J Cell Biol. 2002;157(1):63–77.
29. Maxfield FR, van Meer G. Cholesterol, the central lipid of mammalian cells. Curr Opin Cell Biol. 2010;22(4):422–9.

30. Urbani L, Simoni RD. Cholesterol and vesicular stomatitis virus G protein take separate routes from the endoplasmic reticulum to the plasma membrane. J Biol Chem. 1990;265(4):1919–23.
31. Baumann NA, Sullivan DP, Ohvo-Rekila H, Simonot C, Pottekat A, Klaassen Z, et al. Transport of newly synthesized sterol to the sterol-enriched plasma membrane occurs via nonvesicular equilibration. Biochemistry. 2005;44(15):5816–26.
32. Goldstein JL, Brown MS. The LDL receptor. Arterioscler Thromb Vasc Biol. 2009;29(4):431–8.
33. Maxfield FR, McGraw TE. Endocytic recycling. Nat Rev Mol Cell Biol. 2004;5(2):121–32.
34. D'Angelo G, Vicinanza M, De Matteis MA. Lipid-transfer proteins in biosynthetic pathways. Curr Opin Cell Biol. 2008;20(4):360–70.
35. Mesmin B, Antonny B, Drin G. Insights into the mechanisms of sterol transport between organelles. Cell Mol Life Sci. 2013 Sept;70(18):3405–21.
36. Clark BJ. The mammalian START domain protein family in lipid transport in health and disease. J Endocrinol. 2012;212(3):257–75.
37. Prinz WA. Non-vesicular sterol transport in cells. Prog Lipid Res. 2007;46(6):297–314.
38. Beh CT, McMaster CR, Kozminski KG, Menon AK. A detour for yeast oxysterol binding proteins. J Biol Chem. 2012;287(14):11481–8.
39. Tsujishita Y, Hurley JH. Structure and lipid transport mechanism of a StAR-related domain. Nat Struct Biol. 2000;7(5):408–14.
40. Iyer LM, Koonin EV, Aravind L. Adaptations of the helix-grip fold for ligand binding and catalysis in the START domain superfamily. Proteins-Struct Funct Genet. 2001;43(2):134–44.
41. Ponting CP, Aravind L. START: a lipid binding domain in StAR, HD-ZIP and signalling proteins. Trends Biochem Sci. 1999;24(4):130–2.
42. Soccio RE, Breslow JL. StAR-related lipid transfer (START) proteins: mediators of intracellular lipid metabolism. J Biol Chem. 2003;278(25):22183–6.
43. Romanowski MJ, Soccio RE, Breslow JL, Burley SK. Crystal structure of the *Mus musculus* cholesterol-regulated START protein 4 (StarD4) containing a StAR-related lipid transfer domain. Proc Natl Acad Sci U S A. 2002;99(10):6949–54.
44. Lavigne P, Najmanivich R, Lehoux JG. Mammalian StAR-related lipid transfer (START) domains with specificity for cholesterol: structural conservation and mechanism of reversible binding. Sub-cell Biochem. 2010;51:425–37.
45. Murcia M, Faraldo-Gomez JD, Maxfield FR, Roux B. Modeling the structure of the StART domains of MLN64 and StAR proteins in complex with cholesterol. J Lipid Res. 2006;47(12):2614–30.
46. Bose HS, Whittal RM, Baldwin MA, Miller WL. The active form of the steroidogenic acute regulatory protein, StAR, appears to be a molten globule. Proc Natl Acad Sci U S A. 1999;96(13):7250–5.
47. Friedland N, Liou HL, Lobel P, Stock AM. Structure of a cholesterol-binding protein deficient in Niemann-Pick type C2 disease. Proc Natl Acad Sci U S A. 2003;100(5):2512–7.
48. Im YJ, Raychaudhuri S, Prinz WA, Hurley JH. Structural mechanism for sterol sensing and transport by OSBP-related proteins. Nature. 2005;437(7055):154–8.
49. Roderick SL, Chan WW, Agate DS, Olsen LR, Vetting MW, Rajashankar KR, et al. Structure of human phosphatidylcholine transfer protein in complex with its ligand. Nat Struct Biol. 2002;9(7):507–11.
50. Kudo N, Kumagai K, Tomishige N, Yamaji T, Wakatsuki S, Nishijima M, et al. Structural basis for specific lipid recognition by CERT responsible for nonvesicular trafficking of ceramide. Proc Natl Acad Sci U S A. 2008;105(2):488–93.
51. Soccio RE, Adams RM, Romanowski MJ, Sehayek E, Burley SK, Breslow JL. The cholesterol-regulated StarD4 gene encodes a StAR-related lipid transfer protein with two closely related homologues, StarD5 and StarD6. Proc Natl Acad Sci U S A. 2002;99(10):6943–8.
52. Rodriguez-Agudo D, Calderon-Dominguez M, Ren S, Marques D, Redford K, Medina-Torres MA, et al. Subcellular localization and regulation of StarD4 protein in macrophages and fibroblasts. Biochim Biophys Acta. 2011;1811(10):597–606.

53. Soccio RE, Adams RM, Maxwell KN, Breslow JL. Differential gene regulation of StarD4 and StarD5 cholesterol transfer proteins. Activation of StarD4 by sterol regulatory element-binding protein-2 and StarD5 by endoplasmic reticulum stress. J Biol Chem. 2005;280(19):19410–8.
54. Leventis PA, Grinstein S. The distribution and function of phosphatidylserine in cellular membranes. Annu Rev Biophys. 2010;39:407–27.
55. Fridriksson EK, Shipkova PA, Sheets ED, Holowka D, Baird B, McLafferty FW. Quantitative analysis of phospholipids in functionally important membrane domains from RBL-2H3 mast cells using tandem high-resolution mass spectrometry. Biochemistry. 1999;38(25):8056–63.
56. Bose M, Whittal RM, Miller WL, Bose HS. Steroidogenic activity of StAR requires contact with mitochondrial VDAC1 and phosphate carrier protein. J Biol Chem. 2008;283(14):8837–45.
57. Yamada S, Yamaguchi T, Hosoda A, Iwawaki T, Kohno K. Regulation of human STARD4 gene expression under endoplasmic reticulum stress. Biochem Biophys Res Commun. 2006;343(4):1079–85.
58. Riegelhaupt JJ, Waase MP, Garbarino J, Cruz DE, Breslow JL. Targeted disruption of steroidogenic acute regulatory protein D4 leads to modest weight reduction and minor alterations in lipid metabolism. J Lipid Res. 2010;51(5):1134–43.
59. Rodriguez-Agudo D, Ren S, Hylemon PB, Redford K, Natarajan R, Del Castillo A, et al. Human StarD5, a cytosolic StAR-related lipid binding protein. J Lipid Res. 2005;46(8):1615–23.
60. Rodriguez-Agudo D, Ren S, Hylemon PB, Montanez R, Redford K, Natarajan R, et al. Localization of StarD5 cholesterol binding protein. J Lipid Res. 2006;47(6):1168–75.
61. Bose HS, Whittal RM, Ran Y, Bose M, Baker BY, Miller WL. StAR-like activity and molten globule behavior of StARD6, a male germ-line protein. Biochemistry. 2008;47(8):2277–88.
62. Chen YC, Meier RK, Zheng S, Khundmiri SJ, Tseng MT, Lederer ED, et al. Steroidogenic acute regulatory-related lipid transfer domain protein 5 localization and regulation in renal tubules. Am J Physiol Renal Physiol. 2009;297(2):F380–8.
63. Letourneau D, Lorin A, Lefebvre A, Cabana J, Lavigne P, Lehoux JG. Thermodynamic and solution state NMR characterization of the binding of secondary and conjugated bile acids to STARD5. Biochim Biophys Acta. 2013;1831(11):1589–99.
64. Borthwick F, Allen AM, Taylor JM, Graham A. Overexpression of STARD3 in human monocyte/macrophages induces an anti-atherogenic lipid phenotype. Clin Sci. 2010;119(7):265–72.
65. Tsai YC, Weissman AM. The unfolded protein response, degradation from endoplasmic reticulum and cancer. Genes Cancer. 2010;1(7):764–78.
66. Golas A, Malek P, Piasecka M, Styrna J. Sperm mitochondria diaphorase activity–a gene mapping study of recombinant inbred strains of mice. Int J Dev Biol. 2010;54(4):667–73.
67. Alpy F, Stoeckel ME, Dierich A, Escola JM, Wendling C, Chenard MP, et al. The steroidogenic acute regulatory protein homolog MLN64, a late endosomal cholesterol-binding protein. J Biol Chem. 2001;276(6):4261–9.
68. Miller WL. StAR search–what we know about how the steroidogenic acute regulatory protein mediates mitochondrial cholesterol import. Mol Endocrinol. 2007;21(3):589–601.
69. Rone MB, Fan J, Papadopoulos V. Cholesterol transport in steroid biosynthesis: role of protein-protein interactions and implications in disease states. Biochim Biophys Acta. 2009;1791(7):646–58.
70. Hauet T, Yao ZX, Bose HS, Wall CT, Han Z, Li W, et al. Peripheral-type benzodiazepine receptor-mediated action of steroidogenic acute regulatory protein on cholesterol entry into leydig cell mitochondria. Mol Endocrinol. 2005;19(2):540–54.
71. Tomasetto C, Regnier C, Moog-Lutz C, Mattei MG, Chenard MP, Lidereau R, et al. Identification of four novel human genes amplified and overexpressed in breast carcinoma and localized to the q11-q21.3 region of chromosome 17. Genomics. 1995;28(3):367–76.
72. Alpy F, Tomasetto C. Give lipids a START: the StAR-related lipid transfer (START) domain in mammals. J Cell Sci. 2005;118(Pt 13):2791–801.

73. Strauss JF 3rd, Kishida T, Christenson LK, Fujimoto T, Hiroi H. START domain proteins and the intracellular trafficking of cholesterol in steroidogenic cells. Mol Cell Endocrinol. 2003;202(1–2):59–65.
74. Wang ML, Motamed M, Infante RE, Abi-Mosleh L, Kwon HJ, Brown MS, et al. Identification of surface residues on Niemann-Pick C2 essential for hydrophobic handoff of cholesterol to NPC1 in lysosomes. Cell Metab. 2010;12(2):166–73.
75. Zhang M, Liu P, Dwyer NK, Christenson LK, Fujimoto T, Martinez F, et al. MLN64 mediates mobilization of lysosomal cholesterol to steroidogenic mitochondria. J Biol Chem. 2002;277(36):33300–10.
76. Kishida T, Kostetskii I, Zhang Z, Martinez F, Liu P, Walkley SU, et al. Targeted mutation of the MLN64 START domain causes only modest alterations in cellular sterol metabolism. J Biol Chem. 2004;279(18):19276–85.
77. Kanno K, Wu MK, Scapa EF, Roderick SL, Cohen DE. Structure and function of phosphatidylcholine transfer protein (PC-TP)/StarD2. Biochim Biophys Acta. 2007;1771(6):654–62.
78. Hanada K, Kumagai K, Tomishige N, Yamaji T. CERT-mediated trafficking of ceramide. Biochim Biophys Acta. 2009;1791(7):684–91.
79. Wirtz KW, Devaux PF, Bienvenue A. Phosphatidylcholine exchange protein catalyzes the net transfer of phosphatidylcholine to model membranes. Biochemistry. 1980;19(14):3395–9.
80. Johnson LW, Zilversmit DB. Catalytic properties of phospholipid exchange protein from bovine heart. Biochim Biophys Acta. 1975;375(2):165–75.
81. Wustner D, Mukherjee S, Maxfield FR, Muller P, Herrmann A. Vesicular and nonvesicular transport of phosphatidylcholine in polarized HepG2 cells. Traffic. 2001;2(4):277–96.
82. Durand S, Angeletti S, Genti-Raimondi S. GTT1/StarD7, a novel phosphatidylcholine transfer protein-like highly expressed in gestational trophoblastic tumour: cloning and characterization. Placenta. 2004;25(1):37–44.
83. Horibata Y, Sugimoto H. StarD7 mediates the intracellular trafficking of phosphatidylcholine to mitochondria. J Biol Chem. 2010;285(10):7358–65.
84. Olayioye MA, Hoffmann P, Pomorski T, Armes J, Simpson RJ, Kemp BE, et al. The phosphoprotein StarD10 is overexpressed in breast cancer and cooperates with ErbB receptors in cellular transformation. Cancer Res. 2004;64(10):3538–44.
85. Olayioye MA, Vehring S, Muller P, Herrmann A, Schiller J, Thiele C, et al. StarD10, a START domain protein overexpressed in breast cancer, functions as a phospholipid transfer protein. J Biol Chem. 2005;280(29):27436–42.
86. Hanada K, Kumagai K, Yasuda S, Miura Y, Kawano M, Fukasawa M, et al. Molecular machinery for non-vesicular trafficking of ceramide. Nature. 2003;426(6968):803–9.
87. van Meer G, Holthuis JC. Sphingolipid transport in eukaryotic cells. Biochim Biophys Acta. 2000;1486(1):145–70.

Index

B. J. Clark, D. M. Stocco (eds.), *Cholesterol Transporters of the START Domain Protein Family in Health and Disease,* DOI 10.1007/978-1-4939-1112-7,

MIX
Papier aus verantwortungsvollen Quellen
Paper from responsible sources
FSC® C105338

If you have any concerns about our products,
you can contact us on
ProductSafety@springernature.com

In case Publisher is established outside the EU,
the EU authorized representative is:
Springer Nature Customer Service Center GmbH
Europaplatz 3, 69115 Heidelberg, Germany

Printed by Libri Plureos GmbH
in Hamburg, Germany